LECTURES ON THE THEORY OF INTEGRATION

Series in Real Analysis Volume 1

LECTURES ON THE THEORY OF INTEGRATION

Ralph Henstock
University of Ulster (Coleraine)

World Scientific
Singapore • New Jersey • Hong Kong

Published by

World Scientific Publishing Co. Pte. Ltd.
P.O. Box 128, Farrer Road, Singapore 9128

U. S. A. office: World Scientific Publishing Co., Inc.
687 Hartwell Street, Teaneck NJ 07666, USA

LECTURES ON THE THEORY OF INTEGRATION

ISBN 9971-50-450-2
9971-50-451-0 (pbk)

Printed in Singapore by General Printing Services Pte. Ltd.

CONTENTS

PREFACE

People turn to integration for a variety of reasons and needs, and these can be roughly classified into four kinds. First there are appropriate integral formulae for the evaluation of volumes of solid figures, areas on flat and curved surfaces, and the lengths of certain curves, together with measures of aggregate attributes such as mass, moments of inertia, charge given the charge density, total magnetism given a magnetic density, and so on. Secondly, when solving differential and integral equations in the sciences and elsewhere, the evaluation of integrals is usually vital. Next, continuous linear functionals and self-adjoint (hypermaximal) linear operators over spaces of functions are represented in spectral theory and elsewhere by Stieltjes-type integrals. Finally, integrals are used for distributions in statistics, and these are in turn often used to approximate certain sums. For example, binomial sums are approximated by using suitable normal (or Gaussian) integrals.

Students usually begin with the calculus inverse of the derivative, the indefinite integral of Newton, and then turn to the definite integral of Leibnitz, Cauchy, Riemann and Darboux, to name the principal mathematicians involved. These simple integrals satisfy many; much is done in physics and applied mathematics by using the calculus alone, without recourse to more modern ideas. However, it is becoming increasingly clear to scientists in general, especially those interested in statistics, statistical physics, or quantum theory, that such a limitation of the pure mathematical tools is a restriction, and that the use of a more powerful integration process will result in a greater ease of application.

Lesbesgue integration began in a famous paper of 1902, has several useful and interesting properties, and is more powerful in some directions than calculus integration. But a student of its theory endures a complete change of ideas. One approach needs a theory of measure to construct the integral. In another approach the integral is defined directly as a continuous linear functional. But, whatever the approach, the theory is formidable and many find it difficult to understand. Yet the *evaluation* of almost all Lebesgue integrals uses calculus methods.

This curious paradoxical situation often irritated the author when researching in Lebesgue and allied integration theory from 1943 to 1958. Further, even in one dimension it is difficult to estimate a basic element of Lebesgue theory, the measureof the union of open neighbourhoods of known lengths of members of an everywhere dense sequence of points, even when the lengths and points have easy formulae. There is much overlapping of neighbourhoods.

A useful tool for mathematicians and scientists needing advanced integration theory would be a method combining the ideas of the calculus indefinite integral and the Riemann definite integral in such a way that Lebesgue properties can be proved easily, so that the student would not have such a radically fresh start as is needed for Lebesgue integration. Before the 1950's it would have seemed to be an impossible dream. But independently J. Kurzweil (1957) and the author (1955, 1961) arrived at the same construction, a simple extension of the construction of the Riemann integral. The resulting integral includes the Lebesgue and Denjoy-Perron integrals and is the subject of this book. It is one of a whole family of integrals definable by using the Riemann sums over divisions with a limit process that is stronger than the limit as the mesh tends to 0. The theory has advanced considerably in the 25 years since writing the first book, which is now out of print. Comparison of that pioneering book with the present book shows how proofs have simplified considerably. Further, conditions have been found that are both necessary and sufficient for a sequence of integrable functions to have an integrable pointwise limit, and for the limit to be taken inside the integral sign. The discrete variable can be replaced by a continuous variable, so that necessary and sufficient conditions are found for differentiation and integration under the integral sign. Thus Sections 11, part of 13, 14, 15 in part, and 18 are new, while in Section 19 the Carathéodory theory of ordinary differential equations is improved.

I wish to express my appreciation to Ms. Terri Moss for the excellent typescript, and to many colleagues and students who have checked the lecture notes and have helped to produce a better theory.

R. Henstock

0 Prerequisites

We begin with the usual notation of set theory. A collection of objects is sometimes called a set and sometimes a family, these two terms being interchangeable. If an object x is a member of a set X we write $x \in X$. Sing (x) is the set whose only member is x . If every member of X is also a member of a set Y we write $X \subseteq Y$ and $Y \supseteq X$. If X and Y have the same members, $X \subseteq Y$ and $Y \subseteq X$, we regard X and Y as being the same set and write $X = Y$. Otherwise we write $X \neq Y$ and say that X,Y are distinct. If $X \subseteq Y$, $X \neq Y$, we write $X \subset Y$, $Y \supset X$. If M is a family of sets X we write,

$$\bigcup_M X, \quad \bigcap_M X$$

for the union and intersection, respectively, of sets X of M. The first is the set of all objects each of which lies in at least one of the sets X of M. The second is the set of all objects each of which lies in all sets of M. Usually M is omitted from the symbols and there are other slight variations. For instance, if M is the collection of X_j for $j = 1,2,\ldots$, we write

$$\bigcup_{j=1}^{\infty} X_j, \quad \bigcap_{j=1}^{\infty} X_j.$$

Note that these definitions do not need any idea of a limit. The union and intersection of two sets X_1, X_2 are for simplicity written as $X_1 \cup X_2$, $X_1 \cap X_2$, respectively. Two sets with no common member, so that the intersection is the empty set, $\emptyset$, are called disjoint. If each pair of sets from a family M of sets is disjoint, we say that the sets of M are mutually disjoint. If X is a set of real numbers the complement $\setminus X$ of X is the set of all real numbers that do not lie in X. Thus $X \cap \setminus X$ is empty and $X \cup \setminus X$ is the set of all real numbers. More generally, $X \setminus Y$ denotes the set of all objects in X that are not in Y. If (X_j) is a sequence of sets then we define

$$\limsup_{j\to\infty} X_j = \bigcap_{m=1}^{\infty} \bigcup_{j=m}^{\infty} X_j, \quad \liminf_{j\to\infty} X_j = \bigcup_{m=1}^{\infty} \bigcap_{j=m}^{\infty} X_j,$$

and if they are the same set we write it as

$$\lim_{j\to\infty} X_j.$$

There is still no idea of a limit in these definitions, even though lim occurs in the names. If $X_1 \subseteq X_2 \subseteq \ldots \subseteq X_j \subseteq \ldots$ then $\lim_j X_j = \bigcup_{j=1}^{\infty} X_j$. We say that (X_j) is monotone increasing. On the other hand, if $X_1 \supseteq X_2 \supseteq \ldots \supseteq X_j \supseteq \ldots$ then $\lim_j X_j = \bigcap_{j=1}^{\infty} X_j$. We say that (X_j) is monotone decreasing. A relation T between the objects x of a set X and the objects y of a set Y, is called a one-one correspondence if T satisfies the following requirements.

(0.1) If $x \in X$ there is a $y \in Y$ such that xTy;

(0.2) if there is a $z \in Y$ such that xTz, then $z = y$;

(0.3) if $y \in Y$, there is an $x \in X$ with xTy;

(0.4) if there is a $w \in X$ with wTy then $w = x$.

We also say that X,Y are similar. Using this idea we could define the positive integers as cardinal numbers, and then proceed to define the real and complex number systems. Instead we assume that these systems are already given and will only use the idea of similarity to define countability. A set X is countable if it is similar to the set of all positive integers. Then a subset of a countable set is finite or countable. Also a finite or countable union of finite or countable sets is finite or countable, which is proved by Cantor's diagonal process. But the set of all real numbers in $0 \leq x < 1$ is not countable. (For each x in the set can be written as a decimal in the scale of 10, say, $x = 0.x_1x_2\ldots x_n\ldots$, where x_n takes an integer value from 0 to 9. Some real numbers have a dual representation, for example, 0.50000... = 0.4999... . To remove this duality we make the convention that no decimal has $x_n = 9$ for all n greater than some integer. If (x_j) is a sequence of decimals, with $x_j = 0.x_{j1}x_{j2}x_{j3}\ldots x_{jn}\ldots$ we can define a decimal $y = 0.y_1y_2y_3\ldots y_n\ldots$ by the rule

$$y_n = x_{nn} + 1 \ (0 \leq x_{nm} \leq 7), \ y_n = 1 \ (x_{nn} = 8,9),$$

so that $y_n \neq 9$. Then y differs from each x_j in at least the jth. decimal place and $0 \leq y < 1$. Thus no sequence can contain all x in $0 \leq x < 1$.)

The rationals (ratios of integers) are countable, for they are a distinct subsequence of the sequence

$$0,1/1,-1/1,2/1,-2/1,1/2,-1/2,3/1,-3/1,2/2,-2/2,1/3,-1/3,4/1,\ldots$$

Or put j/k in sequence by writing j and k as integers in the scale of 10, separated by the /, regard / as the integer 10 in the scale of 11, interpret j and k now as integers in the scale of 11, and we have an integer in the scale of 11. Such an integer cannot be given by any other rational. Thus countability follows, and then the uncountability of the irrationals (those real numbers that are not rational).

In a small part of the book we need a little elementary topology. An open sphere in n-dimensional Euclidean space R^n with centre $x = (x_1,\ldots,x_n)$ and radius $r > 0$, is the set of all $y = (y_1,\ldots,y_n)$ satisfying

$$\sum_{j=1}^{n} (y_j - x_j)^2 < r^2.$$

A set G in R^n is open if each point of G is the centre of a sphere wholly contained in G. A union and a finite intersection of open sets are open. A point x is the limit point of a set X in R^n if every sphere, centre x, contains a point of X other than x. An isolated point of X is a point of X that is not a limit point. The set X' of all limit points of X is called the derived set of X, and $X \cup X'$ is X^c, the closure of X. A set F that contains all its limit points, is called closed. The letter F is taken from fermé, French for 'closed'. A set X is dense-in-itself if every point of X is a limit point of X ($X \subseteq X'$). A set P that is closed and dense-in-itself ($P = P'$) is called perfect, and hence the P. The complement of an open set is closed, the complement of a closed set is open. An intersection and a finite union of closed sets are closed.

A set M is dense in a set X if $M \subseteq X \subseteq M'$. On the real line all points x of X for which, for some $\varepsilon > 0$, there is an interval $(x-\varepsilon, x)$ or an interval $(x, x+\varepsilon)$, or both, free from points of X (i.e. the set Y of points of X isolated on one or both sides) is a finite or countable set.

LECTURES ON THE THEORY OF INTEGRATION

CHAPTER 1 - INTRODUCTION

1. The Riemann And Riemann-Darboux Integrals

The Greeks used geometry to compute many simple areas, giving rise to the method of exhaustions of Eudoxus (c. 408-355 BC) and Archimedes (c. 287-212 BC). In this first crude limit process a sequence of non-overlapping triangles is fitted into the area to exhaust it. Thus they found the areas of the circle and sections of the parabola. But they did not define a general non-negative polynomial over a given range, nor even graphs, and so could not compute the area under a curve given by a non-negative polynomial.

Graphs were introduced in the 17th. century, possibly by R. Descartes (1596-1650) in 1619, Christiaan Huygens mentioned them in a letter to his brother Lodewijk on 21st November 1669. The calculus was also invented and used by I. Newton (1642-1727) and G.W. Leibnitz (1646-1716) at that time.

However, setting the scene for modern integration, we begin with the definite integral on the real line of G.F.B. Riemann (1826-1866) in 1854 (published 1868) as modified by J.G. Darboux (1842-1917) in 1875. For real numbers $a < b$ and the closed interval $a \leq x \leq b$ or $[a,b]$, let $f : [a,b] \to R$. We partition $[a,b]$ into smaller intervals by points

$$(1.1) \qquad a = u_0 < u_1 < u_2 < \ldots < u_{m-1} < u_m = b.$$

Its *mesh* is the maximum of $u_j - u_{j-1}$ for $j = 1,2,\ldots,m$. When $f \geq 0$ in $[a,b]$, the smallest rectangle with base $[u_{j-1}, u_j]$ that encloses the part of the graph of $y = f(x)$ with x in the base, has height the *least upper bound or supremum of* f there,

$$M_j = \sup \{f(x) : u_{j-1} \leq x \leq u_j\}.$$

The collection of m rectangles, for $j = 1,2,\ldots,m$, has area

$$U = \sum_{j=1}^{m} M_j(u_j - u_{j-1}),$$

the *upper Darboux sum* for the partition. Similarly the rectangle with base $[u_{j-1}, u_j]$ that lies just below the graph, has height the *greatest lower bound or infimum*

$$m_j = \inf \{f(x) : u_{j-1} \leq x \leq u_j\}.$$

The area of this collection of rectangles, for $j = 1,2,\ldots,m$, is

$$L = \sum_{j=1}^{m} m_j(u_j - u_{j-1}),$$

the *lower Darboux sum* for the partition. If f is sometimes negative in [a,b] the m_j, M_j, L,U can still be defined in this way, but of course the geometric picture changes slightly. Clearly $U \geq L$. If these sums tend to the same limit I for arbitrary partitions (1.1) of [a,b] as the mesh tends to 0, we say that the *Riemann-Darboux integral of* f exists over [a,b] with value I. Then U and L have to be finite for some partition, and

(1.2) *if* f *is Riemann-Darboux integrable over* [a,b] *then* f *is bounded there.*

Riemann's 1854 definition (published 1868) is different. For f and (1.1) as before, let x_j be an arbitrary point of $[u_{j-1}, u_j]$ $(j=1,\ldots,n)$ with

(1.3) $$s = \sum_{j=1}^{m} f(x_j)(u_j - u_{j-1}), = (E)\ \Sigma f(x)(v-u).$$

The division E is the collection of $([u_{j-1}, u_j], x_j)(j = 1,\ldots,m)$ used in (1.3), its *mesh* (E) is the mesh of (1.1), and $(E)\ \Sigma$ denotes summation over E. If every such s tends to a fixed number I as *mesh* $(E) \to 0$, we say that the *Riemann integral of* f *exists* over [a,b] with value I, and we write $I = (R)\int_{[a,b]} f dx$. More precisely:

(1.4) *given* $\varepsilon > 0$, *a* $\delta > 0$ *depending on* ε *is such that if mesh* $(E) < \delta$ *then* $|s-I| < \varepsilon$.

If L and U tend to the same number I as *mesh* $(E) \to 0$, so does s since $L \leq s \leq U$. Conversely, in (1.3) with a fixed partition (1.1), for $\varepsilon > 0$, there are $x_1, \ldots, x_m$ with

$$M_j \geq f(x_j) > M_j - \varepsilon (j = 1, \ldots, m), \quad U \geq s > U - \varepsilon(b-a),$$

b-a being finite. For another choice of $x_1, \ldots, x_m$,

$$m_j \leq f(x_j) < m_j + \varepsilon \; (j = 1, \ldots, n), \quad L \leq s \leq L + \varepsilon(b-a).$$

As $\varepsilon > 0$ is arbitrary, if $s \to I$ as *mesh* $(E) \to 0$, then $L \to I$, $U \to I$. Thus by (1.2),

(1.5) *if f is real valued, the Riemann and Riemann-Darboux definitions yield the same results, and an integrable function is bounded.*

A.L. Cauchy (1789-1857) gave a similar definition in 1821 in which $x_j = u_j (j = 1, \ldots, m)$.

Riemann (1868) art. 5, shows that

(1.6) *the necessary and sufficient condition for the Riemann integral of a bounded function to exist over* [a,b], *is that the total length of the subintervals for which the oscillation is greater than any fixed positive number, is arbitrarily small.*

Those already knowing Lebesgue theory can note that here appears the idea of a Riemann-integrable function being bounded, and continuous almost everywhere, though not expressed in the Lebesgue notation.

We can take $x_j = u_j$ or $x_j = u_{j-1}$, a choice for each j in 1,2,...,m. For if $u_{j-1} < x_j < u_j$ we can replace $[u_{j-1}, u_j]$ by $[u_{j-1}, x_j]$ and $[x_j, v_j]$, while the sum is not altered since

$$f(x_j)(u_j - u_{j-1}) = f(x_j)(x_j - u_{j-1}) + f(x_j)(u_j - x_j).$$

Ex. 1.1. Evaluate

$$\lim_{j\to\infty} (1/(j^2+1) + 2/(j^2+4) + \ldots + j/(j^2+j^2))$$

(University of Ulster 1986, M 112).

Hint: write the expression inside the main brackets as a sum of $j^{-1}(rj^{-1})(1 + (rj^{-1})^2)^{-1}$, for $r = 1,\ldots,j$, a Riemann sum to integrate $x(1+x^2)^{-1}$ from 0 to 1.

Ex. 1.2. Similarly find

$$\lim_{j\to\infty} \sum_{k=1}^{j} (2k-1)^5.2^{-5}j^{-6}.$$

(New University of Ulster 1975, M 111).

Ex. 1.3. If $0 < r < 1$ and if $f(0) = 0$ and $f(x) = r^m(r^m < x \leq r^{m-1})$, $m = 1,2,\ldots$, prove that f is Riemann integrable over $0 \leq x \leq 1$ with value $r/(1+r)$, even though f has an infinity of discontinuities.

(New University of Ulster 1972, M 112)

Ex. 1.4. Let (s_j) be a not necessarily monotone sequence of points in $[0,1]$ with infimum s, and put

$$f(x) = (s_j < x) \sum s_j.j^{-2},\ f(x) = 0\ (x \leq s).$$

Prove that the Riemann integral of f exists over [0,1] with value

$$\sum_{j=1}^{\infty} s_j(1-s_j)j^{-2}.$$

(New University of Ulster 1972, M 213)

2. Modifications Using The Mesh And The Refinement Of Partitions

The Riemann-Darboux definition depends on there being a reasonable order in the space of values of f, and can be extended to functions with values in lattices. But it can only be extended to complex-valued functions by integrating the real and imaginary parts separately, which causes complications. The Riemann definition, however, is far more versatile, it can be used for functions with values in lattices, the complex plane, the set of quaternions, and even a topological group. Again, if for real-valued or complex-valued functions we replace sums by products, we have a Riemann product integral, see J.D. Dollard and C.N. Friedman (1979) which gives many references.

T.J. Stieltjes (1856-1894) in 1894 modified (1.3), using another function g and replacing $f(x)(v-u)$ by $f(x)\{g(v)-g(u)\}$, and replacing s by

$$(2.1) \qquad s_g = \sum_{j=1}^{m} f(x_j)\{g(u_j)-g(u_{j-1})\} = (E)\ \Sigma\ f(x)\{g(v)-g(u)\}.$$

The resulting integral as the mesh tends to 0, is called the *Riemann-Stieltjes integral* and is written $(RS)\int_{[a,b]} f\,dg$. If $g(v)-g(u)$ always remains of the same sign we can define a *Riemann-Darboux-Stieltjes integral* when necessary.

J.C. Burkill (1924) replaced $f(x)(v-u)$ by a function $h(u,v)$ of the interval from u to v, or an *interval function*, obtaining the *Burkill integral*.

The author uses an interval-point function $h([u,v),x)$, and in Henstock (1963), p. 26, one finds $\{h_\ell, h_r\}$ where $h_\ell(u,v) = h([u,v),v)$, $h_r(u,v) = h([u,v),u)$.

If a real- or complex-valued function g, defined on $[a,b]$, is such that for some positive number M,

$$\sum_{j=1}^{m} |g(u_j) - g(u_{j-1})| \leq M$$

for every partition (1.1) of $[a,b]$, we say that g *is of bounded variation on* $[a,b]$. The smallest value of M is the *variation* $\mathrm{var}(g;[a,b])$ *of* g. Then g is bounded, since

$$|g(x)| \leq |g(a)| + |g(x) - g(a)| + |g(b) - g(x)| \leq |g(a)| + M.$$

<u>Theorem 2.1</u>: *If* f *is continuous and* g *of bounded variation on* $[a,b]$, *then* f *is Riemann-Stieltjes integrable relative to* g *on* $[a,b]$.

Proof: By uniformity of continuity, given $\varepsilon > 0$, there is a $\delta > 0$, such that

(2.2) $$|f(v) - f(u)| < \varepsilon \quad (a \le u < v \le b,\ v-u < 2\delta).$$

Let E be the division in (1.3) with *mesh* (E) the mesh of (1.1), and let E_p' be the special division consisting of $([v_{j-1},v_j),v_j)$ $(j = 1,\dots,p)$ obtained by using $v_j = a + (b-a)j/p$ $(j = 0,1,\dots,p)$. We consider the intersections of every overlapping pair of intervals from (1.1) and E_p and use (2.2), taking *mesh* $(E) < \delta$, $b-a < p\delta$. If

$$[u_{j-1},u_j] \cap [v_{k-1},v_k] = [s,t], \text{ then } |x_j - v_k| < 2\delta,$$

$$|f(x_j)\{g(t)-g(s)\}-f(v_k)\{g(t)-g(s)\}| = |f(x_j)-f(v_k)|\,|g(t)-g(s)| \le \varepsilon|g(t)-g(s)|,$$

$$|(E)\Sigma\, f(x)\{g(v)-g(u)\} - (E_p)\Sigma f(x)\{g(v)-g(u)\}| \le \varepsilon\, \Sigma\, |g(t)-g(s)| \le \varepsilon\, M.$$

Taking $E = E_q$ with $q\delta >$ b-a we see that

$$((E_p)\ \Sigma\ f(x)\{g(v)-g(u)\})$$

is a fundamental sequence and so tends to a limit I, and then for arbitrary E with mesh less than δ, and $p \to \infty$, the result follows from

$$|(E)\ \Sigma\ f(x)\{g(v)-g(u)\} - I| \le \varepsilon M.$$

Theorem 2.2: *As mesh* $(E) \to 0$ *let*

$$(E)\ \Sigma\ |(f(v)-f(u))(g(v)-g(u))| \to 0.$$

If f *is Riemann-Stieltjes integrable relative to* g, *then* g *is Riemann-Stieltjes integrable relative to* f *and*

(2.3) $$(RS)\int_{[a,b]} f dg + (RS)\int_{[a,b]} g df = f(b)g(b) - f(a)g(a).$$

This is the formula for integrating Riemann-Stieltjes integrals by parts.

Proof: The following is an identity.

$$(2.4)\qquad g(x)\{f(v)-f(u)\} + f(x)\{g(v)-g(u)\} = f(v)g(v)-f(u)g(u) + \{f(x)-f(u)\}\{g(x)-g(u)\} - \{f(v)-f(x)\}\{g(v)-g(x)\}.$$

In (2.4), all sums over divisions of the first two terms on the right are equal to

$$f(b)g(b) - f(a)g(a),$$

sums of the last two products tend to 0 as *mesh* $(E) \to 0$, and the sum over E of the second term on the left tends to the Riemann-Stieltjes integral of f over [a,b] relative to g. Hence g is Riemann-Stieltjes integrable over [a,b] relative to f, with (2.3).

<u>Theorem 2.3</u>: *If* g *is of bounded variation and* f *continuous on* [a,b], *then* g *is Riemann-Stieltjes integrable relative to* f *on* [a,b] *with* (2.3).

<u>Proof</u>: From (2.2) and an arbitrary division E of [a,b] with *mesh* $< \delta$,

$$(E)\,\Sigma\,|f(v)-f(u)|\cdot|g(v)-g(u)| \leq (E)\,\Sigma\,\varepsilon|g(v)-g(u)| \leq \varepsilon M$$

and so tends to 0 as *mesh* $(E) \to 0$. Hence the result follows by Theorems 2.1, 2.2.

Note that if f is continuous but not of bounded variation, with g of bounded variation, one cannot define a Radon or Lebesgue-Stieltjes integral of g relative to f, even though the Riemann-Stieltjes integral exists.

<u>Theorem 2.4</u>: f *is Riemann-Stieltjes integrable relative to itself and*

$$(2.5)\qquad (RS)\int_{[a,b]} f\,df = \tfrac{1}{2}f^2(b) - \tfrac{1}{2}f^2(a),$$

if as mesh $(E) \to 0$,

$$(2.6)\qquad (E)\,\Sigma\,|f(v)-f(u)|^2 \to 0.$$

The converse is true when f *is real-valued.*

Proof: In (2.4) with $g = f$, the two terms on the left are equal and the sum of $f^2(v)-f^2(u)$ over E is $f^2(b)-f^2(a)$. Hence (2.5). For the converse take $x = v$ and $g = f$ in (2.4). The next theorem shows that with continuity, the boundedness of the variation can in some sense be split up between f and g.

Theorem 2.5: *Let* f,g *be continuous in* [a,b] *with respective moduli of continuity* $\lambda(h)$, $\mu(h)$, (*i.e.* $\lambda(h) = \sup\{|f(y)-f(x)|:|y-x| \le h, (x,y) \subseteq (a,b)\}$, *and similarly for* $\mu(h)$.) *If*

$$M(h) \equiv \int_{[0,h]} \frac{\lambda(t)\mu(t)}{t^2}\, dt$$

converges for small $h > 0$, *then* (RS) $\int_{[a,b]} f dg$ *exists.*

Proof: Writing E_p^* for the division E'_{2p} of Theorem 2.1, we prove the existence of

$$\lim_{p\to\infty} H_p, = H, \text{ say, where } H_p = (E_p^*) \sum f(x)\{g(v)-g(u)\}.$$

For $h = v-u > 0$ let $[u,v)$ have midpoint y. Then

$$|f(v)\{g(v)-g(u)\}-f(y)\{g(y)-g(u)\}-f(v)\{g(v)-g(y)\}| =$$

$$|f(v)-f(y)||g(y)-g(u)| \le \lambda(\tfrac{1}{2}h)\mu(\tfrac{1}{2}h).$$

Repeating the bisection till $[u,v)$ has a partition P of 2^p equal intervals,

$$(2.7) \qquad |f(v)\{g(v)-g(u)\}-(P)\sum f(t)\{g(t)-g(s)\}| \le$$

$$\sum_{j=1}^{p} 2^{j-1}\lambda(h2^{-j})\mu(h2^{-j}) \le 2(v-u)M(v-u),$$

since for $x > 0$ and then $x = h2^{-j}$,

$$(2.8) \qquad \int_x^{2x} \frac{\lambda(t)\mu(t)}{t^2}\, dt \ge \frac{\lambda(x)\mu(x).x}{(2x)^2}$$

Hence for $q > p$,

$$|H_p - H_q| \leq 2(b-a)M((b-a)2^{-p}) \to 0$$

as $p \to \infty$, and (H_p) is a Cauchy or fundamental sequence and so has a limit, say H, with

$$(2.9) \qquad |H_p - H| \leq 2(b-a)M((b-a)2^{-p}).$$

For P a partition of m equal intervals with $2^{p-1} < m < 2^p$ we extend $[u,v)$ to $[u,w)$ by adding $2^p - m$ intervals each of length h/m, with f,g constant on $[v,w]$. Then the new length is

$$h + (2^p - m)h/m = 2^p h/m < 2h, \text{ and } f(v)\{g(v)-g(u)\} = f(w)\{g(w)-g(u)\}.$$

For some $\delta > 0$ with $4M(2\delta) < \varepsilon$, and $v-u < \delta$, (2.7) becomes

$$(2.10) \qquad |f(v)\{g(v)-g(u)\} - (P)\ \Sigma f(t)\{g(t)-g(s)\}| \leq 4(v-u)M(2(v-u)) < \varepsilon(v-u).$$

Next, let $I_1, \ldots, I_r$ be a partition P^* of $[a,b)$ of mesh $< \delta/3$, and by continuity of f and g, let $\eta > 0$ be such that

$$|f(v)\{g(v)-g(u)\} - f(t)\{g(t)-g(s)\}| < \varepsilon(b-a)/r \ (|v-t| < \eta,\ |u-s| < \eta).$$

For $(b-a)2^{-p} < \eta$ let $a = u_0 < u_1 < \ldots < u_m = b$ be the partition from E_p^*. With each I_j associate the interval $[u_s, u_t) = J_j$ whose end-points are the u's that are nearest to the left and right end-points of $I_j (j = 1, \ldots, r)$. If $s = t$, J_j disappears. Then

$$|(P^*)\ \Sigma\ f(v)\{g(v)-g(u)\} - \sum_1^r f(u_t)\{g(u_t)-g(u_s)\}| < \varepsilon(b-a).$$

As $u_t - u_s \leq 3\mathit{mesh}\ (P^*) < \delta$, dividing up those $[u_s, u_t)$ with $t > s+1$, (2.10) and (2.9) give

$$|\sum_1^r f(u_t)\{g(u_t)-g(u_s)\} - H_p| < \varepsilon(b-a), \quad |H_p - H| < \varepsilon(b-a),$$

$$|(P^*)\ \Sigma f(v)\{g(v)-g(u)\} - H| < 3\varepsilon(b-a),$$

$$(P^*)\ \Sigma f(v)\{g(v)-g(u)\} \to H\ (\mathit{mesh}\ (P^*) \to 0).$$

Finally we change v to x in $u \leq x \leq v$. From (2.8), if $0 < x \leq \delta$,

$$4M(2x) < \varepsilon, \quad \lambda(x)\mu(x) < \varepsilon x,$$

so that if $(I_j, x_j)(j = 1,\ldots,r)$ form a division E of $[a,b)$ of mesh $< \delta/3$,

$$|(E)\,\Sigma f(x)\{g(v)-g(u)\}-(P^*)\,\Sigma f(v)\{g(v)-g(u)\}| \leq (P^*)\Sigma\lambda(v-u)\mu(v-u)$$

$$< \varepsilon\ (P^*)\,\Sigma\,(v-u) = \varepsilon(b-a).$$

Corollary: The proof of the theorem shows that, given $\varepsilon > 0$, there is a $\delta > 0$, depending on ε and the functions λ,μ alone, and independent of the particular f,g with respective moduli of continuity not greater than λ,μ, such that

$$|(E)\ \Sigma\ f(x)\{g(v)-g(u)\}-(RS)\int_{[a,b]} fdg| < \varepsilon \quad (\mathit{mesh}\ (E) < \delta).$$

Theorem 2.6: *If* $g:[a,b] \to R$ *or* C, *is of bounded variation in* $[a,b]$, g *has an at most countable number of discontinuities* (a_j) *in* $[a,b]$, *each one of which can be approached from above (except if* $a_j = b$*) and below (except if* $a_j = a$*) by points where* g *is continuous. Further, there exist*

(2.11) $$g(x-) = \lim_{u\to x-} g(u)\ (a < x \leq b),\ g(x+) = \lim_{u\to x+} g(u)\ (a \leq x < b),$$

(2.12) $$\sum_{j=1}^{\infty} \{|g(a_j)-g(a_j-)| + |g(a_j+)-g(a_j)|\} \leq M = \mathrm{var}(g:[a,b]).$$

(2.13) $g_c(x)$ *is continuous in* $[a,b]$, *where*

$$g_c(a) = g(a),\ g_c(x) = g(x) - \Sigma_1\{g(a_j)-g(a_j-)\}-\Sigma_2\{g(a_j+)-g(a_j)\},$$

and Σ_1 *is over all* a_j *in* $a < a_j \leq x$, Σ_2 *is over all* a_j *with* $a \leq a_j < x$.

(2.14) *If* f *is continuous, then, except for obvious changes when* $a_j = a$ *or* b,

$$(RS)\int_{[a,b]} fdg = (RS)\int_{[a,b]} fdg_c + \sum_{j=1}^{\infty} f(a_j)\{g(a_j+)-g(a_j-)\}.$$

Proof: For $[u,v)$ the set of all x in $u \leq x < v$ let $([u_j,v_j))$ be a sequence of mutually disjoint intervals in $[a,b)$. Then, for each m,

$$[a,b] \setminus \bigcup_{j=1}^{m} [u_j,v_j)$$

is a finite number of intervals, so that by their partitioning,

$$(2.15) \qquad \sum_{j=1}^{m} |g(v_j) - g(u_j)| \leq M \ (m = 1,2,\ldots)$$

and the infinite series is convergent. First take $u_1 = a$, $u_j = v_{j-1}$ $(j = 2,3,\ldots)$, $v_j \to c \leq b$. Then

$$G \equiv \lim_{j\to\infty} g(v_j) = g(a) + \sum_{j=1}^{\infty} \{g(v_j) - g(v_{j-1})\}$$

exists as an absolutely convergent series. For (w_j) also monotone increasing to c, $g(w_j)$ similarly tends to a limit, say G^*. Rearranging (v_j) and (w_j) as a single monotone increasing sequence we find $G^* = G$ and the limit of $g(v_j)$ is independent of the particular sequence (v_j), and so $g(c-)$ exists. Similarly for $g(c+)$ in $a \leq c < b$. Then to prove (2.12) we take m points a_j in $[a,b]$ and $2m$ mutually disjoint intervals $[u_j,a_j)$, $[a_j,v_j)$ in (2.15) (omitting an interval if $a_j = a$ or b), and so obtain (2.12) with m replacing ∞. Thus at most mM points have $|g(a_j)-g(a_j-)| \geq 1/m$, and at most mM points have $|g(a_j+)-g(a_j)| \geq 1/m$, and we can put such points for $m = 1,2,\ldots$ in sequence as (a_j). Hence the discontinuities are at most countable, and we have (2.12). The series in (2.13) are thus absolutely convergent and that result follows. For (2.14) we use Theorem 2.1 and only need prove that if f is continuous in $[a,b]$, (a_j) a sequence of points in (a,b), and $\sum_{j=1}^{\infty} c_j$ is absolutely convergent then

$$(2.16) \qquad \sum_{j=1}^{\infty} c_j f(a_j) = (RS)\int_{[a,b]} f dh, \quad h(x) = \Sigma\{c_j: a < a_j \leq x\},$$

and similarly for $a \leq a_j < x$ replacing $a < a_j \leq x$ in the definition of h. To prove (2.16), let the division E consist of $([u_{k-1},u_k),x_k)(k=1,2,\ldots,r)$ with mesh $(E) < \delta$, where

$$|f(x) - f(y)| < \varepsilon (a \le x < y \le b, \ |x-y| < \delta).$$

The result follows from

$$\left|\sum_{j=1}^{\infty} c_j f(a_j) - \sum_{k=1}^{r} f(x_k) \{h(u_k) - h(u_{k-1})\}\right| =$$

$$\left|\sum_{k=1}^{r} \Sigma \{c_j[f(a_j)-f(x_k)] : u_{k-1} < a_j \le u_k\}\right| \le \varepsilon \sum_{j=1}^{\infty} |c_k|.$$

Finally the points of continuity are dense in [a,b] since the discontinuities are countable and the points in each interval are not countable, so a point of continuity lies in every interval.

For $c > 0$ let $\phi : [0,c] \to R$ be monotone increasing. Note that $+\infty$ is not in R, so that we are implying that ϕ is finite. Then ϕ is of bounded variation as every difference is non-negative, and the moduli of differences add up to $\phi(c) - \phi(0)$. For convenience we assume that

$$\phi(x) = \phi(x-) \quad (0 < x \le c)$$

and we define $\psi: [\phi(0), \phi(c)] \to [0,c]$ by

$$\psi(y) = \inf \{x: 0 \le x \le c, \ \phi(x) \ge y\}.$$

We say that ψ is the *inverse function of* ϕ, *in Young's sense.*

Theorem 2.7: *For* ϕ,ψ *as given above,* ψ *is finite and monotone increasing and*

(2.17) $\psi(y) = \psi(y-) \ (\phi(0) < y \le \phi(c))$,

(2.18) $\psi(\phi(x)) = \inf \{z : \phi(z) = \phi(x), \ 0 \le z \le x\} \quad (0 \le x \le c)$.

(2.19) *If also* ϕ *is continuous at* $\psi(y)$ *and* $\phi(0) < y < \phi(c)$ *then* $\phi(\psi(y)) = y$.

Proof: If $\phi(0) \le v < y \le \phi(c)$ then $\psi(v) \le \psi(y)$ by the monotonicity of ϕ. If for fixed k and some y in $\phi(0) < y \le \phi(c)$, $\psi(y-) < k < \psi(y)$, then for every $v < y$ we have $\psi(v) < k$, $v \leqq \phi(k) < y$. But this is false for $v = \frac{1}{2}\{\phi(k) + y\} < y$. Hence (2.17). For (2.18), $\psi(\phi(x)) = \inf \{z: \phi(z) \geqq \phi(x), 0 \leqq z \leqq c\}$ while one value of z is x. Hence the result. For (2.19), from $x > \psi(y), \phi(x) \geqq y$ and $\phi(\psi(y)+) \geqq y$. From $x < \psi(y)$, $\phi(x) < y$ and $\phi(\psi(y)-) \leqq y$. By continuity of ϕ we have (2.19).

<u>Theorem 2.8</u>: *For* ϕ, ψ *as before, and* $0 \leqq a < b \leqq c$, *then*

$$(2.20) \qquad (R)\int_{[a,b]} \phi dx + (R)\int_{[\phi(a),y]} \psi dy = by - a\phi(a) \quad (\phi(b) \leqq y \leqq \phi(b+)).$$

<u>Proof</u>: As ϕ, ψ are both monotone increasing and finite, they are of bounded variation, and by Theorem 2.3 the two integrals in (2.20) exist. It follows that we can use any two suitable sequences of divisions with mesh tending to 0, to prove (2.20). Let

$$a = u_0 < u_1 < \ldots < u_m = b$$

be a partition of [a,b] with mesh less than ε. For each u_j that lies in an interval $(v,w] \subseteq [a,b]$ in which $\phi(w) = \phi(v) +$ but $\phi(x) < \phi(v)$ for all $x < v$, we replace u_j by v. The number of intervals may then drop as several u_j may correspond to the same v. For the sake of the notation we then change the other u_j to v_k for some $k \leqq j$. This change may increase the mesh to beyond ε, but the sum over $[v,u_j]$, for the u_j that are moved, is trivially $\phi(w)(u_j-v)$, so that the increase in mesh can be brought to be less than ε by introducing extra partition points on the interval. We now consider the partition

$$(2.21) \qquad \phi(a) = \phi(v_0) < \phi(v_1) < \ldots < \phi(v_p) = b \leqq y.$$

By Theorem 2.7 (2.18),

$$\psi(\phi(v_k)) = v_k \quad (k = 1,\ldots,p-1),$$

while from the definition of ψ, $\psi(y) = b$. Thus we obtain

$$\sum_{k=1}^{p} \phi(v_{k-1})(v_k-v_{k-1}) + \sum_{k=1}^{p-1} \psi(\phi(v_k))\{\phi(v_k)-\phi(v_{k-1})\}+\psi(y)(y-\phi(v_{k-1}))$$

$$= \sum_{k=1}^{p} \phi(v_{k-1})(v_k-v_{k-1}) + \sum_{k=1}^{p-1} v_k\{\phi(v_k)-\phi(v_{k-1})\}+ b(y-\phi(v_{p-1}))=by-a\phi(a)$$

after much cancellation. The mesh of (2.21) need not tend to 0, because of the discontinuities of ϕ. But by Theorem 2.6 (2.13), $\phi = \phi_c + \phi_s$ where ϕ_c is continuous and ϕ_s is a jump function (*fonction des sauts*), a sum of $\phi(s+)-\phi(s)$ at the jumps s of ϕ. The differences of ϕ_c can be made arbitrarily small by taking the partition of [a,b] of arbitrarily small mesh, while the effect of each jump $\phi(s+)-\phi(s)$ is to introduce an interval on which ψ is constant at the value s. Extra division points can be taken on a finite number of such intervals without affecting the sum over $[\phi(a),y]$, so that the final division is of arbitrarily small mesh. Hence (2.20) follows.

Theorem 2.9: *In Theorem* 2.8, *if* $0 \leq a \leq c$, $0 = \phi(0) \leq b \leq \phi(c)$, *then*

(2.22) $\quad \Phi(a) + \Psi(b) \geq ab$ *where* $\Phi(a) \equiv (R)\int_{[0,a]} \phi(x)dx \ (a > 0),\ \Phi(0) = 0,$

$$\Psi(b) \equiv (R)\int_{[0,b]} \psi(y)dy \ (b > 0),\ \Psi(0) = 0.$$

Equality occurs in (2.22) *if, and only if,*

(2.23) $\quad \phi(a) \leq b \leq \phi(a+).$

Proof: Since $\phi(0) = 0$, ψ is defined in $[0,\phi(c)]$, and the results are obvious if a = 0 or b = 0, using Theorem 2.8 (2.20) for the case of equality. If $c \geq x > \phi(a+)$, then by definition of ψ, $\psi(x) > a$. Hence, using (2.20) repeatedly, if $a > 0$, $b > 0$, and $b > \phi(a+)$,

$$\Phi(a)+\Psi(b) = \Phi(a)+\Psi(\phi(a+)) + (R)\int_{[\phi(a+),b]} \psi dy > a\phi(a+)+a(b-\phi(a+)) = ab.$$

If $0 < b < \phi(a)$, there is a u in $0 \leq u < a$ with

$$\phi(u) \leq b \leq \phi(u+),\ \phi(x) > b \ (u < x \leq a),\ \Phi(a)+\Psi(b) = \Phi(u)+\Psi(b)+(R)\int_{[u,a]} \phi\, dx$$
$$> ub + b(a-u) = ab.$$

Hence if equality in (2.22) then (2.23) is true, while if (2.23) is true, (2.20) gives equality in (2.22).

Theorem 2.10: *Let* X *be a set of real numbers and, for each* $u \geq 0$, *let* $g(u,x):X \to R$ *be monotone increasing in* X *with modulus bounded by* M *independent of* n. *Then there are a sequence* $u_j \to \infty$ *and a monotone increasing function* g *in* X, *such that*

$$\lim_{j\to\infty} g(u_j,x) = g(x) \quad (x \in X).$$

Proof: We begin with a lemma.

Lemma 2.11: *Let* $g(u)$ *be a function of positive real numbers with values in* $[-M,M]$, *for some* $M > 0$. *Then for* $t = \limsup_{u\to\infty} g(u)$ *there is a sequence* (u_j) *tending to infinity with* $g(u_j) \to t$.

Proof of Lemma: The set $S(v)$, of values of $g(u)$ for all $u \geq v$, has a supremum, say $s(v)$, and if $w > v$,

$$S(w) \subseteq S(v), \quad -M \leq s(w) \leq s(v) \leq M. \quad \textit{Let } t = \inf s(v) \ (\textit{all } v).$$

Then t exists in $[-M,M]$ and is the upper limit of $g(u)$ as $u \to \infty$. Let

$$u_1 < u_2 < \ldots < u_{j-1},\ u_k \geq k \ (k = 1,\ldots,j-1)$$

be defined inductively. Then as s is monotone decreasing there are a $v_j \geq \max(u_{j-1},j)$ *with* $t \leq s(v_j) < t + j^{-1}$, and as $s(v_j)$ is a supremum, a $u_j \geq v_j$ with

$$t-j^{-1} \leq s(v_j)-j^{-1} < g(u_j) \leq s(v_j) < t + j^{-1}.$$

Thus we continue the induction and so prove the lemma.

Proof of Theorem: Let (x_j) be a sequence of points dense in X, such that the sequence includes all isolated points of X and all points of X that are isolated on one side. Such points can be put in sequence. Then all other points of X are approached on both sides by members of (x_j). Applying the lemma to $g(u,x_1)$, there is a sequence (u_{j1}) tending to infinity, such that $g(u_{j1},x_1)$ tends to a limit, say $g(x_1)$. Applying the lemma to $g(u_{j1},x_2)$, there is a subsequence (u_{j2}) of (u_{j1}) with $g(u_{j2},x_2)$ tending to a limit, say $g(x_2)$, while also $g(u_{j2},x_1)$ tends to $g(x_1)$. This is the start of an inductive proof and we have $\lim_{j\to\infty} g(u_{jk})$ existing, for each k, with value $g(x_k)$, say. Then the diagonal sequence (u_{jj}) is a subsequence of (u_{jk}) for $j \geq k$, so that for each fixed k, $g(u_{jj},x_k) \to g(x_k)$ as $j \to \infty$.

As $g(u,x)$ is monotone increasing in x for each fixed $u \geq 0$, $g(x_k)$ is also monotone increasing and lies in $[-M,M]$. For a point $y \in X$ that is not an x_k, write $g(y-)$ for the limit of $g(x_k)$ as x_k tends to y from below. Then as $g(u,x)$ is monotone increasing in x, $g(u_{jj},y) \geq g(u_{jj},x_k)$, $\liminf_{j\to\infty} g(u_{jj},y) \geq g(x_k)$, $\liminf_{j\to\infty} g(u_{jj},y) \geq g(y-)$. Writing $g(y+)$ for the limit of $g(x_k)$ as x_k tends to y from above, we have similarly, $\limsup g(u_{jj},y) \leq g(y+)$.

Thus when $g(y-) = g(y+)$, $g(u_{jj},y)$ tends to the common value. As $g(x_k)$ is monotone increasing, the points y where $g(y-) \neq g(y+)$ are at most countable and so can be put in a sequence (y_k). We can now go through the same process again for that sequence and prove the theorem with a subsequence of (u_{jj}).

<u>Corollary</u>: *If in Theorem* 2.10, $g(u,x)$, *as a function of* $x \in X$, *is of variation bounded by* M *independent of* u, *there are a sequence* (u_j) *tending to infinity and a function* g *of* $x \in X$, *of variation bounded by* M *with*

$$\lim_{j\to\infty} \{g(u_j,x) - g(u_j,y)\} = g(x) - g(y) \quad (x,y \in X).$$

<u>Proof</u>: Let $a,x,y \in X$ and $a \leq x < y$. Then as $[x,y]$ is equal to $[a,y]\setminus[a,x)$,

$$|g(u,y)-g(u,x)| \leq \mathrm{var}(g(u,\cdot);[a,y])-\mathrm{var}(g(u,\cdot);[a,x]).$$

It follows that h_1,h_2 are monotone increasing on the right of a, where

$$h_1(u,x) = \mathrm{var}(g(u,.);[a,x]) + g(u,x),\ h_2(u,x) = \mathrm{var}(g(u,\cdot);[a,x])-g(u,x),$$

and as the variation is not greater than M,

$$|h_j(u,x) - h_j(u,a)| \leq 2M \ (j = 1,2).$$

Applying Theorem 2.10, the result is true for $h_1(u,x)-h_1(u,a)$ and a sequence (v_j), and then the result is true for $h_2(u,x)-h_2(u,a)$ and a subsequence (w_j) of (v_j). As

$$g(u,x)-g(u,a) = \tfrac{1}{2}(h_1-h_2),$$

the result is true for $g(u,x)-g(u,a)$ in $x \geq a$. For a further subsequence it is true in $x \leq a$ and so true for all x. Then the Corollary is proved since

$$g(u,x)-g(u,y) = g(u,x)-g(u,a)-g(u,y)+g(u,a).$$

Note that $g(u,a)$ could take arbitrary finite values without violating the condition on the variation, and this is why we have to use differences.

The limit process that uses the mesh tending to 0, causes difficulties for Riemann-Stieltjes integration when the two functions have common discontinuities. E.H. Moore (1915) changed this limit process, supposing that for some number I and each $\varepsilon > 0$, there is a special partition

$$(2.24) \qquad a = v_0 < v_1 < \dots < v_p = b$$

with the property that if the partition

$$(1.1) \qquad a = u_0 < u_1 < u_2 < \dots < u_{m-1} < u_m = b$$

is a subpartition of (2.24) (i.e. every v_k is a u_j for some j) with $u_{j-1} \leq x_j \leq u_j$ $(j = 1,\dots,m)$ then the sum s_g of (2.1) satisfies $|s_g-I| < \varepsilon$. Such a partition (1.1) is called a *refinement* of the partition (2.24), and the integral so defined is called a *refinement integral* or *σ-integral*. This can be generalized to partitions of arbitrary sets, and the present one-dimensional case is a little stronger than the Riemann-Stieltjes case using the mesh tending to 0. For if the mesh of (2.24) is less than some $\delta > 0$, so is the mesh of any refinement. Thus

(2.25) *if the Riemann (or Riemann-Stieltjes or Burkill) integral exists, so does the corresponding refinement integral.*

The improvement made by the refinement integral is that for certain points c in $a < c < b$, sums using $[u,v]$ with $a \leq u < c < v \leq b$ are ignored and only corresponding sums using $[u,c]$ and $[c,v]$ are used. For sums (2.1),

$$g(v) - g(u) = g(v) - g(c) + g(c) - g(u).$$

However, $f(x)$ multiplies the differences, and the ranges of x are different in the ignored sums from those in the accepted sums. If f is continuous or g continuous, the changes of sum tend to 0 as $u-v \to 0$. But if f and g are discontinuous at the same point, sometimes the refinement integral exists but the Riemann-Stieltjes integral does not. For example,

$$f(x) = \begin{cases} 0(0 \leq x \leq 1) \\ 1(1 < x \leq 2) \end{cases}, \quad g(x) = \begin{cases} 0(0 \leq x < 1) \\ 1(1 \leq x \leq 2) \end{cases}$$

show the difference. For if 1 is not an end-point of an interval $[u,v]$ of a partition of $[0,2]$, so that $u < 1 < v$, then $g(v) - g(u) = 1$ and $f(x)$ takes both values 0,1. For all other intervals of the partition the difference of values of g is 0. Thus sums take values 0 and 1, whatever the size of mesh, and the Riemann-Stieltjes integral cannot exist. However, we can split the partition at 1 in the case of the refinement integral, and for $[u,1]$ and $[1,v]$ we have the respective values

$$g(1)-g(u) = 1,\ f(x) = 0,\ f(x)\{g(1)-g(u)\} = 0, \text{ and}$$
$$g(v)-g(1) = 0,\ f(x)\{g(v)-g(1)\} = 0.$$

Hence the refinement integral exists and is 0.

An almost trivial change renders the refinement integral useless. The integral of f relative to itself over $[0,2]$ does not exist, for the only non-zero term is that for $[1,v]$, and $f(x)$ can be $0(x = 1)$ or $1(x > 1)$, while $f(v)-f(1) = 1$, whatever the size of $v-1 > 0$. More generally, the refinement integral does not exist when f and g have discontinuities on the same side of the same point, the side lying in the interval of integration.

Again, let the rationals be put in a sequence of distinct points as (r_m), and let $f_j(x)$ be 1 for $x = r_1,\ldots,r_j$, and 0 otherwise. Then over any interval the Riemann integral of f_j is 0, but the Riemann integral and the corresponding refinement integral of the limit function as $j \to \infty$, do not exist. Thus, although the refinement integral is an improvement of the Riemann-Stieltjes integral, the improvement is marginal and questions involving limits under the integral sign cannot be tackled easily, in the way that the Lebesgue and Radon integrals can do, and so we turn to another integral in the next section.

Other problems in integration theory are connected with unbounded functions and unbounded ranges of integration. Neither problem can be tackled using the Riemann or Riemann-Stieltjes or refinement integrals alone, but it was early seen that the continuity of the integral could be used to extend the definitions to cover some of the problems. For example, the derivative of $x^{\frac{1}{2}}$ is $\frac{1}{2}x^{-\frac{1}{2}}$ except at $x = 0$, where there is no finite limit for the derivative. But it was thought legitimate to integrate $\frac{1}{2}x^{-\frac{1}{2}}$ to $x^{\frac{1}{2}}$ + constant even when one end of the range of integration is $x = 0$.

A.L. Cauchy systematised earlier work, developing what we now call the *Cauchy limit*, using the length function and not the Stieltjes difference of values of g. If there is a singularity of the integrand f at a point a, in the sense that the Riemann integral does not exist over $[a,b]$ for any $b > a$, and if the limit of the integral over $[u,b]$ exists as $u \to a+$, we define the value of the integral to be that limit. Similarly if the singularity is at b. If at $c \in (a,b)$ we deal with $[a,c]$ and $[c,b]$ separately. The integral over $[a,\infty)$ is similarly defined to be the limit of the integral over $[a,b]$ as $b \to +\infty$; for $(-\infty,b]$ we let $a \to -\infty$; and for $(-\infty, +\infty)$ we take the limit as $-a$ and b tend independently to $+\infty$. This is different from the limit of the integral over $[-b,b]$ as b tends to $+\infty$, which limit occurs in complex variable theory. If the limit exists with a,b independent, so does the limit with $a = -b$. But if the last exists, there may be a balance between an integral going to $+\infty$ on one side and to $-\infty$ on the other, which balance would be broken when a and b are independent, and the other integral would not exist.

Given these definitions, various tests were devised to ensure that the Cauchy limits exist. We begin with the case of a singularity at a in the interval $[a,b]$.

Theorem 2.12: Let

(2.26) $(R)\int_{[u,b]} f dx$ *exist for each* u *in* $a < u < b$.

(2.27) *For fixed numbers* q, M, v *with* $0 < q < 1$, $M > 0$, $v \in (a,b)$

$$|f(x)|\,(x-a)^q \leq M \quad (a < x \leq v).$$

Proof: *For* $a < u < w < v < b$,

$$\left|(R)\int_{[u,b]} f dx - (R)\int_{[w,b]} f dx\right| = \left|(R)\int_{[u,w]} f dx\right| \leq$$

$$(R)\int_{[u,w]} M(x-a)^{-q} dx = [M(x-a)^{1-q}/(1-q)]_u^w \leq M(w-a)^{1-q}/(1-q) \to 0$$

as $w \to a+$.

Hence the result. It is a rather crude test.

Theorem 2.13: *(Comparison test) Let the point function* $g > 0$, *with numbers* M,N *in* $0 < M < N$.

(2.28) *For a* $v \in [a,b]$ *let* $Mg(x) \leq |f(x)| \leq Ng(x)$ $(a < x \leq v)$, *with* (2.26) *for* g *replacing* f. *If* (2.26) *is true for* f *and so for* $|f|$, *the integrals of* $|f|$ *and* g *converge or diverge together.*

Proof: Again we can use the general principle of convergence.

These two tests are for absolute convergence, which gives convergence. For more delicate tests we turn to analogues of the Abel and Dirichlet tests for infinite series.

Lemma 2.14: *If* (2.26) *is true for* f, *and if* g *is of bounded variation in* $[u,b]$, *then* (2.26) *is true for* fg *replacing* f.

Proof: By Theorem 2.3 g is Riemann-Stieltjes integrable relative to the integral of f over [x,b], which is continuous in x, and this Riemann-Stieltjes integral is equal to the Riemann integral of -fg, by the analogue of Theorem 5.7 (5.25) and since f is bounded in [u,b].

Theorem 2.15: (Abel-Hardy). *If the integral in* (2.26) *converges as* $u \to a+$ *and if* g *is of bounded variation in* [a,b], *then the Riemann integral of* fg *over* [u,b] *converges as* $u \to a+$.

Proof: In Theorem 2.2, for F(x) the Riemann integral of f over [x,b], and its continuity,

$$(E)\ \Sigma\,|\Delta F|\ |\Delta g| \leq \varepsilon\ \Sigma\ |\Delta g| \leq \varepsilon M.$$

Hence as in Lemma 2.14,

$$(R)\int_{[u,b]} fg\,dx = (RS)\int_{[u,b]} -g\,dF(x,b) = -0+g(u)F(u,b)$$

$$+\ (RS)\int_{[u,b]} F(x,b)\,dg(x) = g(u)\{F(u,b)-F(a,b)\} + F(a,b)g(b)$$

$$+\ (RS)\int_{[u,b]} \{F(x,b)-F(a,b)\}\,dg.$$

$$|F(x,b)-F(a,b)| < \varepsilon \quad (a < x \leq x_0).$$

Hence, the last Riemann-Stieltjes integral tends to a limit as $a < u < v \leq x_0$,

$$\left|(RS)\int_{[u,b]} - (RS)\int_{[v,b]}\right| = \left|(RS)\int_{[u,v]}\right| \leq \varepsilon\ \mathrm{var}(g;[a,b]).$$

A slight variation on this proof gives the next theorem.

Theorem 2.16: (Dirichlet-Hardy) *If* F *is bounded as* $u \to a+$ *and if* g(x) *is of bounded variation in* [a,b] *and tends to* 0 *as* $x \to a+$, *then the Riemann integral of* fg *over* [u,b] *converges as* $u \to a+$.

These tests can easily be translated into applying when the singularity occurs at b.

Turning to integrals over $[a, +\infty)$ for fixed a, we suppose that

(2.29) *for each fixed* a *and each* $b > a$, f *is integrable on* $[a,b]$ *by Riemann's method or by using Cauchy limits on Riemann integrals.*

Theorem 2.17: *For positive constants* C, p, X, *let* $|f(x)| \leq Cx^{-1-p}$ $(x \geq X)$. *Then the integral of* f *exists over* $[a, +\infty)$.

Proof: For $F(u)$ the integral of f over $[a,u]$, if $a < u < v$, $u > X$, we have

$$|F(v)-F(a)| = \left|\int_{[u,v]} f dx\right| \leq \int_{[u,v]} Cx^{-1-p} dx = C(u^{-p}-v^{-p})/p < Cu^{-p}/p \to 0 \quad (u \to +\infty),$$

giving convergence,

There are a comparison and Abel andDirichlet tests for $[a,\infty)$, of which a statement of the Abel test is given. The others are easy to write by analogy with Theorems 2.13, 2.16.

Theorem 2.18: (Abel-Hardy). *If* $\int_{[a,+\infty)} f dx$ *converges and if* g *is of bounded variation in* $[a.+\infty)$, *then* $\int_{[a,+\infty)} fg dx$ *converges.*

If a series converges, its mth. term tends to 0 as $m \to \infty$. But the function inside an integral convergent over $[a,+\infty)$ need not have that property. For let the graph of f in $[m-m^{-3}, m+m^{-3}]$ be the sloping sides of a triangle, with $f = 0$ at the ends of the interval and with the triangle vertex at (m,m), for $m = 2,3,4,\ldots$, and let f be 0 outside those intervals. Then the intervals are mutually disjoint. The integral of f over the interval with vertex (m,m) is the area m^{-2} of the triangle, and $\sum_{m=2}^{\infty} m^{-2}$ is convergent. Thus as f is non-negative, $(R)\int_{[0,+\infty)} f dx$ is convergent. But $f(m) = m \to \infty$.

However, if $f \geq 0$ and is monotone decreasing, things are more regular.

Theorem 2.19: *Let* $f \geq 0$ *and be monotone decreasing in* $x \geq 1$. *Then*

(2.30) $$\sum_{j=1}^{m} f(j) \geq (R)\int_{[1,m+1]} f(x)dx \geq \sum_{j=2}^{m+1} f(j),$$

(2.31) $0 \leq s_m \leq f(1)-f(m+1)$, *where* $s_m \equiv \sum_{j=1}^{m} f(j)-(R)\int_{[1,m+1]} f(x)dx.$

(2.32) s_m *tends to a limit* p *as* $m \to \infty$, *where* $0 \leq p \leq f(1)-\lim_{x\to\infty} f(x).$

(2.33) *(The integral test). If* $f \geq 0$ *is monotone decreasing in* $x \geq 1$ *then* $\sum_{j=1}^{\infty} f(j)$ *and* $\int_{[1,+\infty)} f(x)dx$ *converge or diverge together.*

Proof: If $j \leq x \leq j+1$ then $f(j) \geq f(x) \geq f(j+1)$. Integrating and taking $j = 1,2,\ldots,m$ we have (2.30) and then (2.31). Further,

$$s_{m+1}-s_m = f(m+1) - (R)\int_{[m+1,m+2]} f(x)dx \geq 0,$$

and (s_m) is monotone increasing, and bounded by (2.31), and so tends to a limit p, giving (2.32). Then (2.33) follows.

Theorem 2.20: (Frullani). *Let* $\int_{[A,B]} f(x)/x\,dx$ *exist by some reasonable definition, for each finite* A,B *in* $0 < A < B$. *Then for* $0 < a < b$,

(2.34) $$\int_{[A,B]} \frac{f(bx) - f(ax)}{x}\,dx = \int_{[aB,bB]} \frac{f(x)}{x}\,dx - \int_{[aA,bA]} \frac{f(x)}{x}\,dx.$$

Thus, to study the convergence of the integral over $[0, +\infty)$ *we can separate the behaviour of* f *for large* x *from the behaviour of* f *for small* $x > 0$.

(2.35) *If there is a constant* C *such that*

$$\int_{[A,+\infty)} \frac{f(x)-C}{x}\,dx$$

exists, then as $B \to \infty$,

(2.36) $$\int_{[aB,bB]} \frac{f(x)}{x}\, dx \to C \log_e \frac{b}{a}.$$

(2.37) *In particular this occurs if*

$$\int_{[A,B]} \{f(x)-C\}dx$$

is bounded or tends to a limit as $B \to \infty$.

(2.38) *If* $f(x) \to C$ *as* $x \to \infty$, (2.36) *holds.*

(2.39) *If there is a constant* K *such that*

$$\int_{[0,B]} \frac{f(x)-K}{x}\, dx$$

is convergent then

(2.40) $$\int_{[aA,bA]} \frac{f(x)}{x}\, dx \to K \log_e \frac{b}{a}\ .$$

(2.41) *Also* (2.40) *occurs if* $f(x) \to K$ *as* $x \to 0+$.

If (2.36) *and* (2.40) *are true then*

(2.42) $$\int_{[0,+\infty)} \frac{f(bx)-f(ax)}{x}\, dx$$

converges to $(C-K)\log_e(b/a)$.

Note that $1/x$ is of bounded variation in $[A,+\infty)$ but not in $[0,A]$, so that the analogue of (2.37) does not occur for (2.39).

Proof: Let the first integral in the statement be denoted by $F(B)-F(A)$. Then for $y = ax$ and $z = bx$,

$$\int_{[A,B]} \frac{f(bx)-f(ax)}{x}\, dx = \int_{[bA,bB]} \frac{f(z)}{z}\, dz - \int_{[aA,aB]} \frac{f(y)}{y}\, dy$$

$$= F(bB) - F(bA) - F(aB) + F(aA) = \{F(bB) - F(aB)\} - \{F(bA)-F(aA)\}$$

giving (2.34). For (2.35), (2.36) is obvious. Also (2.37) gives (2.35) by Abel's or Dirichlet's tests, as $1/x$ is of bounded variation. For (2.38), if x is large enough,

$$\left|\int_{[aB,bB]} \frac{f(x)-C}{x}\, dx\right| \leq \int_{[aB,bB]} \frac{\varepsilon}{x}\, dx = \varepsilon \log_e(b/a),$$

$$\int_{[aB,bB]} \frac{C}{x}\, dx = C \log_e(b/a).$$

(2.39), (2.40) follow (2.35, 2.36), while (2.41) follows (2.38). Then for (2.42) we use (2.34), (2.36), (2.40).

Some examples involve the Young functions ϕ, ψ.

Ex. 2.1. Let (s_m) be an arbitrary sequence of distinct numbers in $(0,1)$, and let

$$\phi(x) = x + (s_m < x) \Sigma\, 2^{-m}.$$

Show that the inverse function ψ is continuous with derivative lying between 0 and 1. Thus ψ is the integral of its derivative.

Ex. 2.2. Let $x = 0 \cdot x_1 x_2 \ldots x_m \ldots$ be a decimal to base 3, i.e. each decimal place x_m can be filled by the integers 0,1,2, but no others. Such decimals represent the closed interval $[0,1]$. Let C be the subset of all such decimals with $x_m \neq 1$ for all m. Show that in a repeated trisection of $[0,1]$ the open middle interval in each trisection is missing from C. (C is called *Cantor's ternary set.*)

For each decimal x in C let y denote the corresponding binary decimal $0, y_1 y_2 \ldots y_m \ldots$ (i.e. decimal to base 2) where $2y_m = x_m (m = 1,2,\ldots)$. Writing $y = \phi(x)$ in C and then taking ϕ constant in each of the open intervals that make up $[0,1] \setminus C$, such that ϕ is continuous in $[0,1]$, then ϕ is monotone increasing from 0 to 1. Find the inverse function ψ of ϕ.

Ex. 2.3. Let ϕ be finite and monotone *decreasing* in $[0,c]$, and let ψ be the inverse function of $-\phi$. Prove that if $0 \leq a < b \leq c$,

$$(R)\int_{[a,b]} \phi\, dx - (R)\int_{[-\phi(a),-\phi(b)]} \psi\, dy = b\phi(b) - a\phi(a).$$

Ex. 2.4. If $c > 0$, $r > 1$, $0 < rx \leq c$, and ϕ is finite and monotone increasing in $[0,c]$, then

$$\Phi(rx) \geq \Phi(x) + (r-1)x\phi(x) \geq r\Phi(x),\ \Phi(x) \equiv (R)\int_{[0,x]} \phi(t)dt.$$

Ex. 2.5. For $\phi(x) = x^{p-1}$ where $p > 1$ is fixed, show that $\psi(y) = y^{1/(p-1)}$, $\Phi(x) = x^p/p$, and $\Psi(y) = y^q/q$ where $q = 1 + 1/(p-1) = p/(p-1)$ and $1/p + 1/q = 1$. Thus $a^p/p + b^q/q \geq ab$ ($a \geq 0$, $b \geq 0$, $p > 1$, $1/p + 1/q = 1$, with equality if and only if $a^p = b^q$.

Ex. 2.6 For $\phi(x) = \begin{cases} 0 (0 \leq x \leq e^{-1}) \\ \log_e x + 1\ (x \geq e^{-1}) \end{cases}$, $\psi(y) \begin{cases} 0 & (y = 0) \\ e^{y-1} & (y > 0) \end{cases}$, $\psi(0+) = e^{-1}$,

$$\Phi(x) = \begin{cases} 0\ (0 \leq x \leq e^{-1}) \\ x\log x + e^{-1} (x > e^{-1}) \end{cases},\ \Psi(y) = e^{y-1} - e^{-1}.$$

Thus

$$a\log_e a + e^{b-1} \geq ab\ (a \geq e^{-1},\ b \geq 0) \tag{2.43}$$

with equality for $b = \log a + 1$. If $0 < a < e^{-1}$, $b \geq 0$, then

$$e^{b-1} - e^{-1} \geq ab.$$

As $a\log_e a$ has a minimum at $a = e^{-1}$, we have (2.43) again, with strict inequality. Hence (2.43) is true for $a > 0$, $b \geq 0$.

Ex. 2.7. $(R)\int_{[u,1]} \log_e x\, dx = [x\log_e x - x]_u^1 = -1 - u\log_e u + u \to -1$ as $u \to 0+$.

Ex. 2.8. $(R)\int_{[u,1]} x\log_e x\, dx = [\tfrac{1}{2}x^2\log_e x - \tfrac{1}{4}x^2]_u^1 = -\tfrac{1}{4} - \tfrac{1}{2}u^2\log_e u + \tfrac{1}{4}u^2 \to -\tfrac{1}{4}$ as $u \to 0+$.

(New University of Ulster, 1978, M 112).

Ex. 2.9. $(R)\int_{[0,u]} (1-x^2)^{-\frac{1}{2}}\, dx = \text{arc sin } u - \text{arc sin } 0 \to \tfrac{1}{2}\pi\ (u \to 1-)$.

Ex. 2.10. $(R)\int_{[u,1]} (\sin(1/x) - (1/x)\cos(1/x))dx = [x\sin(1/x)]_u^1$

$= \sin 1 - u\sin(1/u) \to \sin 1$. For $-1 \leq \sin B \leq +1$.

Ex. 2.11. Consider the convergence and evaluation of

$$(R)\int_{[p,q]} x\{(q-x)(x-p)\}^{-\frac{1}{2}}\,dx \quad (0 < p < q).$$

The use of $x = p\cos^2\theta + q\sin^2\theta$ removes the singularity.

Ex. 2.12 Examine the convergence of

$$(R)\int_{[0,1]} (1-x)^{-\frac{1}{2}}\sin x\,dx.$$

Ex. 2.13. For what values of the constant p are the integrals

$$\int_{[0,1]} x^{-p}\sin x\,dx, \quad \int_{[0,1]} x^{-p}\cos x\,dx,$$

convergent? Note that

$$0 \leq x^{-p}\sin x \leq x^{1-p} \quad (0 < x \leq 1).$$

(New University of Ulster 1983, M 112).

Ex. 2.14. Is the following integral convergent or divergent?

$$\int_{[0,1]} \sqrt{x}.\mathrm{cosec}\,x\,dx.$$

(New University of Ulster, 1973, M 211).

Ex. 2.15. $I = \int_{[0,\frac{1}{2}\pi]} \log_e \sin x\,dx$

is a mos interesting integral. Over the range $(0,\frac{1}{2}\pi)$ the graph of sin x is high .han the chord joining the end-points (0,0) and $(\frac{1}{2}\pi,1)$. Hence $\sin x > (2x)/\pi$, $|\log_e \sin x| = \log_e \mathrm{cosec}\,x \leq \log_e(\pi/(2x))$ which tends to $+\infty$ slower than any negative power of x. Hence convergence from

$$x^{\frac{1}{2}}\log_e \sin x \to 0 \quad (x \to 0+).$$

To evaluate I put $x = \frac{1}{2}\pi - y$, and later, $z = 2x$, $t = \pi - z$, to give

$$I = \int_{[0,\frac{1}{2}\pi]} \log_e \cos y\, dy,\quad 2I = \int_{[0,\frac{1}{2}\pi]} \log_e(\sin x \cos x)\, dx$$

$$= \int_{[0,\frac{1}{2}\pi]} \log_e(\tfrac{1}{2}\sin 2x)dx = -\tfrac{1}{2}\pi\log_e 2 + \tfrac{1}{2}I$$

$$+ \int_{[\frac{1}{2}\pi,\pi]} \log_e \sin z\, dz = -\tfrac{1}{2}\pi \log_e 2 + I,\quad I = -\tfrac{1}{2}\pi \log_e 2.$$

Ex. 2.16. Show that $f:[0,1] \to R$ given by $f(x) = 2^n.n^{-2}$ for $1 - 2^{1-n} \leq x < 1-2^{-n}$, and $f(1) = 0$, is integrable over $[0,1]$ by a Cauchy limit.

(University of Ulster, 1986, M 112).

Ex. 2.17. $\displaystyle\int_{[0,+\infty)} \frac{x^{p-1}}{1+x}\, dx$. Prove that this is convergent if and only if $0 < p < 1$.

(New University of Ulster, 1975, M 211).

Ex. 2.18. Prove that $A = \displaystyle\int_{[0,1]} \frac{\log_e x}{1+x^2}\, dx$, $B = \displaystyle\int_{[1,+\infty)} \frac{\log_e x}{1+x^2}\, dx$ are convergent with $B = -A$. Deduce that if $a > 0$ then

$$\int_{[0,+\infty)} \frac{\log_e x}{a^2+x^2}\, dx = \pi(\log_e a)/(2a).$$

(New University of Ulster, 1978, M 211)

Ex. 2.19. Show that the following is convergent, and evaluate it.

$$\int_{[1,+\infty)} x^{-2} \log_e x\, dx$$

(New University of Ulster, 1977, M 112).

Ex. 2.20. Integrating by parts, show that if $p > 0$ is a constant,

$$\int_{[1,+\infty)} x^{-p}\cos x\, dx \quad\text{and}\quad \int_{[1,+\infty)} x^{-p}\sin x\, dx$$

are convergent. Using Ex. 2.13, find the ranges of p for which the corresponding integrals over $[0,+\infty)$ are convergent.

Ex. 2.21. Using integration by parts prove the convergence of

$$\int_{[2,+\infty)} \frac{\sin x}{\log_e x}\,dx$$

Ex. 2.22. Prove the convergence of the sequence whose mth. term is

$$\sum_{j=1}^{m} \frac{1}{j} - \log_e(m+1)$$

Ex. 2.23. For $0 < A < B$, $0 < a < b$ rearrange the integral

$$\int_{[A,B]} \frac{\cos bx - \cos ax}{x}\,dx$$

in terms of integrals of $\cos x/x$ over two ranges, one range depending only on A and another depending only on B. Using $|1-\cos x| < x^2$ for small x, in the first range and integrating by parts for the *second range,* show that as $A \to 0+$ and $B \to +\infty$, the integral tends to $\log_e(a/b)$. (New University of Ulster, 1977, M 211).

Ex. 2.24. In Ex. 2.20, when p is fixed in $0 < p \leq 1$ the second integral is not absolutely convergent. For

$$(R)\int_{[\pi,m\pi]} \frac{|\sin x|}{x^p}\,dx = \sum_{j=2}^{m} \int_{[0,\pi]} \frac{\sin y}{\{y+(j-1)\pi\}^p}\,dy \geq \sum_{j=2}^{m} \int_{[0,\pi]} \frac{\sin y}{(j\pi)^p}\,dy$$

$$= \frac{2}{\pi^p} \sum_{j=2}^{m} \frac{1}{j^p} \to \infty \quad \text{as } m \to \infty .$$

3. The Calculus Indefinite Integral And The Riemann-Complete Or Generalized Riemann Integral

A still earlier definition of the integral is that of Newton. The *calculus indefinite integral (or Newton integral)* F *of the function* f *over* $[a,b]$, *is any solution of the differential equation*

(3.1) $\quad dy/dx = f(x)$

in $[a,b]$, the derivative being finite at each point of $[a,b]$. Thus, given

$\varepsilon > 0$, there is a $\delta = \delta(x) > 0$, depending on ε and x, such that

(3.2) $$\left|\frac{F(x+p) - F(x)}{p} - f(x)\right| < \varepsilon , \quad |F(x+p) - F(x) - f(x)p| < \varepsilon \; |p|$$

for $a \leqq x \leqq b$, $a \leqq x+p \leqq b$, $0 < |p| < \delta(x)$.

The first point is that the Newton integral can only integrate derivatives, and these have one property of continuous functions. Darboux has shown that

(3.3) *if* $F'(x)$ *exists at all points of* $[a,b]$ *with* $F'(b) \neq F'(a)$, *and if* z *lies strictly between* $F'(a)$ *and* $F'(b)$, *there is an* x *in* $a < x < b$ *with* $F'(x) = z$.

Proof: We can replace $F(x)$ by $\pm(F(x)-zx)$ and so assume that

(3.4) $F'(a) < 0 = z < F'(b)$.

Then the infimum in $[a,b]$ is attained at some point x in $[a,b]$ since F is continuous, and x cannot be a nor b, because of the inequalities in (3.4). Hence $F'(x) = 0$, proving the result. Hence

(3.5) *if* $g(x) = 0(x < 0)$, $g(x) = 1(x \geqq 0)$, *then* g *is not a derivative and does not have a Newton integral, since* g *does not take the value* $\frac{1}{2}$.

Continuous functions on a bounded interval or a compact set are bounded, but not all derivatives are bounded and Riemann integrable. For example, if

(3.6) $$F(x) = x^2\sin(1/x^2) \; (x \neq 0), \; F(0) = 0,$$
$$F'(x) = 2x\sin(1/x^2)-(2/x)\cos(1/x^2)(x \neq 0), \; F'(0) = 0,$$

and F' is unbounded at 0.

There is no standard calculus construction of the Newton integral, but when found, we recognise it by the defining property (3.1). Thus, as usual in the calculus, we have to look at the differentiation process

$DF = dF/dx$ and so the operator D with the following properties. Let F,G be differentiable and let m,p be constants. Then

(3.7) $D(mF + pG) = mDF + pDG,$

(3.8) $D(FG) = (DF)G + F(DG),$

(3.9) $D(F(G)) = (dF/dG)DG,$

(3.10) $Dm = 0.$

Thus the integral, which we can write $F = D^{-1}f$, has the following properties.

(3.11) $D^{-1}(mf + pg) = mD^{-1}f + pD^{-1}g$

for derivatives f,g and constants m,p,

(3.12) $D^{-1}(FDG) = FG - D^{-1}((DF)G),$

the formula for integration by parts, which follows from (3.8). From (3.9) with D_* for d/dG and $h = D_*F$, we have integration by substitution,

(3.13) $D_*^{-1}h = D^{-1}(hDG).$

The more general integral to be defined later, still in some sense obeys (3.11), (3.12), and (3.13). By (3.10), $F = D^{-1}f$ is indeterminate as we can add an arbitrary constant, but as it is constant it cancels when we calculate $F(b) - F(a)$.

(3.14) *The Newton integral of a derivative* f *over* [a,b], *is uniquely defined, apart from an arbitrary constant.*

<u>Proof:</u> The derivative f is given finite for each x in [a,b], so that if $DF = f = DG$ then $D(F-G) = f-f = 0$, for each x in [a,b]. By the mean value theorem, F-G is constant and so $F(b) - F(a) = G(b) - G(a)$.

(3.15) *If* f *is Newton integrable to* F *over* [a,b], *and Riemann integrable to* I *then* $I = F(b) - F(a)$.

Proof: By $D'F = f$ and the mean value theorem for each interval of the partition (1.1) we have a Riemann sum s, using suitable x_j,

$$F(b)-F(a) = \sum_{j=1}^{m} \{F(u_j)-F(u_{j-1})\} = \sum_{j=1}^{m} f(x_j)(u_j-u_{j-1}).$$

As the mesh of (1.1) tends to 0, the last sum tends to I, while the sum remains equal to $F(b)-F(a)$. Hence the result.

From (3.2) we can construct F from the derivative f if we can find a partition (1.1) in which each interval $[u_{j-1},u_j]$ is associated with a point x_j in the interval with $u_j-u_{j-1} < \delta(x_j)$, δ being the function of (3.2) that is defined over $[a,b]$. For

$$0 \leq u_j-x_j \leq u_j-u_{j-1}, \quad 0 \leq x_j-u_{j-1} \leq u_j-u_{j-1},$$

$$|F(u_j) - F(x_j) - f(x_j)(u_j-x_j)| \leq \varepsilon(u_j - x_j),$$

$$|F(x_j) - F(u_{j-1}) - f(x_j)(x_j-u_{j-1})| \leq \varepsilon(x_j-u_{j-1}),$$

$$|F(u_j) - F(u_{j-1}) - f(x_j)(u_j-u_{j-1})| \leq \varepsilon(u_j-u_{j-1}).$$

It follows that

$$(3.16) \qquad |F(b) - F(a) - s| \leq \varepsilon(b-a)$$

where s is as in (1.3), the special Riemann sum for the division E. If also $x_j = u_j$ or $x_j = u_{j-1}$, for $j = 1,2,\ldots,m$, the author has said that E is compatible with δ, or that E is δ-fine. The latter is shorter and will be used, it was first given by McShane, though McShane drops the restriction on x_j and only assumesthat $x_j \in [a,b]$. The function $\delta > 0$ depends on $\varepsilon > 0$, and as $\varepsilon \to 0+$ we would have to let $\delta(x) \to 0+$ at each $x \in [a,b]$, and then by (3.16), s tends to $F(b)-F(a)$. This is the one-dimensional case of the kind of limit we use in this book, and in lectures the author has said at least from 1962 onwards that the limit occurs '*as δ shrinks*'.

If in the division for (1.1), $u_{j-1} < x_j < u_j$ for some j, we can replace $[u_{j-1},u_j]$ by $[u_{j-1},x_j]$ and $[x_j,u_j]$, so that x_j would be at one end of each interval. This will not change the sum s since

$$f(x_j)(u_j-u_{j-1}) = f(x_j)(u_j-x_j) + f(x_j)(x_j-u_{j-1}).$$

It will turn out later that divisions E with $u_{j-1} = x_j$ or $u_j = x_j$ $(j = 1,\dots,m)$ have advantages and better properties than those in which the only restriction is $u_{j-1} \leq x_j \leq u_j$, while we have the same collection of values s.

Using δ-fineness we can now give a formal definition of the Riemann-complete or generalized Riemann integral over the interval from a to b of a function $h([u,v),x)$ of interval-point pairs $([u,v),x)$. First we use intervals $[u,v)$ closed on the left and open on the right, for simplicity of description. If $u < v < w$ then $[u,v)$ and $[v,w)$ are disjoint with union $[u,w)$, of the same type. Using closed intervals in this example, the intervals would have v in common, and we would have to say that the intervals are *non-overlapping*. Next, the *generalized Riemann integral* is a number H with the property that for all δ-fine divisions E of $[a,b)$,

$$(3.17) \qquad |(E)\,\Sigma\, h([u,v),x) - H| < \varepsilon ,$$

where the positive function δ depends on the arbitrarily small $\varepsilon > 0$. We use the notation

$$H = \int_{[a,b)} dh, \quad \text{or} \quad H = \int_{[a,b)} f dx$$

if $h([u,v),x) = f(x)(v-u)$.

One essential requirement is given by the following theorem.

Theorem 3.1: *Given* $\delta(x) > 0$ *in* $[a,b]$, *there is a* δ-*fine division of* $[a,b]$.

Proof: To each $x \in [a,b]$ there corresponds a symmetrical open neighbourhood

$$I(x) \equiv (x-\delta(x),\ x + \delta(x)).$$

By Borel's covering theorem, as a,b are finite a finite number

$I(x_k)$ $(k = 1,\ldots,m)$ *covers* $[a,b]$, i.e. every point of $[a,b]$ lies in at least one of the $I(x_k)$.

(3.18) *We can arrange that each point of* $[a,b]$ *lies in at most two of the* $I(x_k)$.

For let $I(u)$, $I(v)$, $I(w)$ have a common point and $a \leq u < v < w \leq b$. If

$$v-\delta(v) \leq u-\delta(u) \text{ then } \delta(v) \geq \delta(u)+v-u > \delta(u),$$

$$v+\delta(v) > v + \delta(u) > u+\delta(u),\ I(u) \subseteq I(v),$$

and we can omit $I(u)$. Similarly, if $v + \delta(v) \geq w + \delta(w)$ then $I(w) \subseteq I(v)$ and we can omit $I(w)$. Otherwise, as all three intervals have a common point,

$$u - \delta(u) < v-\delta(v) < v + \delta(v) < w+ \delta(w) \text{ implies } I(v) \subseteq I(u) \cup I(w),$$

and we can omit $I(v)$. Successive applications of these results give (3.18), from which we can assume that $a \leq x_1 < x_2 < \ldots < x_m \leq b$ with consecutive $I(x_k)(k = 1,\ldots,m)$ and $a \in I(x_1)$, $b \in I(x_m)$. The intersection

$$(x_j,x_{j+1}) \cap I(x_j) \cap I(x_{j+1})$$

is not empty and so is an open interval, in which we choose a point y_j. Thus $x_j < y_j < x_{j+1}$. We can construct a partition (1.1) with $u_0 = a$, $u_1 = x_1$ (if $x_1 > a$), or otherwise $u_1 = y_1$, while u_2 is y_1 or x_2, respectively, and so on, so that a, b, x_m, x_j, y_j $(j = 1,\ldots,m-1)$ are the division points. By construction the division is δ-fine and the theorem is proved.

Theorem 3.2: *The integral of* h *over* $[a,b)$ *is uniquely defined.*

Proof: Let two numbers H_1, H_2 have the property of H. Then, given $\varepsilon > 0$, there are functions $\delta_k(x) > 0$ such that

(3.19) $\quad |(E_k)\, \Sigma\, h([u,v);x) - H_k| < \varepsilon \ (k = 1,2).$

For $\delta(x) = \min\,(\delta_1(x), \delta_2(x))$, then $\delta(x) > 0$, and by Theorem 3.1 there is a δ-fine division E of $[a,b)$. This division is δ_1-fine and δ_2-fine and so is an E_1 and an E_2, and from (3.19)

$$|H_1 - H_2| = |\{(E)\,\Sigma h - H_2\} - \{(E)\,\Sigma h - H_1\}| < 2\varepsilon.$$

Thus $H_1 = H_2$ as they are independent of $\varepsilon > 0$.

By (3.16) the Newton integral is a generalized Riemann integral; and so is the Riemann integral, which can use constant functions δ. In general the $\delta(x)$ can vary from point to point and, for example, could tend to 0 as x tends to some fixed number w in $[a,b]$ even though $\delta(w) > 0$. A typical construction is as follows:

(3.20) *For all* $x \in [a,b]$ *with* $x \neq w$ *let* $0 < \delta(x) < |x-w|$.

By Theorem 3.1 there is a δ-fine division of $[a,b)$ of the form (1.1) with x_js. If $x_j = u_j \neq w$ then $\delta(x_j) < |u_j - w|$, and neither $[u_{j-1}, u_j]$ nor $[u_j, u_{j+1}]$ can contain w. Thus w is only contained in the closure of one or two intervals of the division with end-point w, and w has to be one of the points of division, with some $x_k = w$, and $[u_{k-1}, w)$, $[w, u_{k+1})$ are intervals of the division, unless w is a or b, when one interval is missing. We can arrange a similar construction for a finite number of w, including a or b or both, or we could use a sequence of points w. See Ex. 4.1 for the n-dimensional case.

Another construction shows that sometimes the Lebesgue integral is also a generalized Riemann integral. Let (X_k) be a sequence of mutually disjoint sets of real numbers, each set being of "measure zero", i.e. for each k and each $\varepsilon > 0$ there is a countable union G of disjoint open intervals I with $G \supseteq X_k$ and $mG < \varepsilon$, where mG is the sum of lengths of the I of G. If X is the union of sets X_k, and if a function f is such that

$$|f(x)| \leq 2^k \ (x \in X_k,\ k = 1,2,\ldots),\quad f(x) = 0 \ (x \in \backslash X),$$

we call f *a null function*.

Theorem 3.3: *The generalized Riemann integral of a null function over* [a,b) *is zero.*

Proof: Given $\varepsilon > 0$, we choose G_k and $\delta(x) > 0$, such that

$$G_k \supseteq X_k, \; mG_k < 4^{-k}\varepsilon, \; \delta(x) \leq 1 \; (x \in \smallsetminus X), \; (x-\delta(x), x+\delta(x)) \subseteq G_k$$

$$(x \in X_k, \; k = 1,2,\ldots).$$

For a δ-fine division E over [a,b), the only nonzero f(x) have $x \in X$, so that

$$|(E) \Sigma f(x)(v-u) - 0| \leq \sum_{k=1}^{\infty} 2^k mG_k < \sum_{k=1}^{\infty} \varepsilon.2^{-k} = \varepsilon.$$

Hence the integral is zero.

A typical null function is the indicator of the rationals, the function that is 1 at each rational and 0 at each irrational. For we can write the rationals p/q as a sequence

$$0/1,\ 1/1,\ -1/1,\ 2/1,\ -2/1,\ 1/2,\ -1/2,\ 3/1,\ -3/1,\ 2/2,\ -2/2,$$

$$1/3,\ -1/3,\ 4/1,\ \ldots$$

in which $|p| + q$ takes the successive values 1 once, 2 twice, 3 four times, 4 six times,... while $|p|$ goes from its maximum down to 1 in each group with constant $|p| + q$, taking $q > 0$. Removing the second and later appearances of each rational, we have a sequence (r_k) that takes each rational once and once only. The intervals

$$(r_k - \varepsilon.2^{-k-1},\ r_k + \varepsilon.2^{-k-1})$$

form a set G with $mG < \varepsilon$, that encloses the sequence (r_k) of rationals. As $\varepsilon > 0$ is arbitrary, the indicator of the rationals is a null function.

The null functions are typical Lebesgue-integrable functions. Assuming a little of Lebesgue theory we can prove that the Lebesgue integral is included in the generalized Riemann integral.

Theorem 3.4: *Let f be a finite real valued function, Lebesgue-integrable over a measurable set M of finite measure, with Lebesgue integral*

$$(L)\int_M f(x)dx = H$$

Then, given $\varepsilon > 0$ *and* $x \in M$, *there is an open set* $G(x)$ *containing* x, *such that whenever*

(3.21) $I_1, I_2, \ldots$ *are disjoint measurable subsets of* M *with*

$$m(M \setminus \bigcup_{j=1}^{\infty} I_j) = 0$$

and $x_1, x_2, \ldots$ *are points satisfying* $x_j \in I_j \subseteq G(x_j)$ $(j = 1,2,\ldots)$, *then*

$$|\sum_{j=1}^{\infty} f(x_j)m(I_j) - H| < \varepsilon.$$

Here the measure of the set X is m(X).

Proof: By the absolute continuity of the Lebesgue integral there is an $\eta > 0$ such that for measurable sets $X \subseteq M$, $m(X) < \eta$ implies

$$(L)\int_X |f|dx < \varepsilon/3.$$

For $\xi = \varepsilon/(3\eta + 3m(M))$ and $k = 0, \pm 1, \pm 2, \ldots$ let

$$X_k = \{x : (k-1)\xi < f(x) \le k\xi\}.$$

Choose for each k an open set $G_k \supseteq X_k$ such that

(3.22) $m(G_k \setminus X_k) < \eta/\{2^{|k|+2}(|k|+1)\}$,

and take $G(x) = G_k$ whenever $x \in X_k$. If (3.22) is satisfied, with $x_j \in X_{k(j)}$, then

$$I_j \subseteq G_{k(j)},\ I_j \setminus X_{k(j)} \subseteq G_{k(j)} \setminus X_{k(j)},\ |\sum_{j=1}^{\infty} f(x_j)m(I_j) - H|$$

$$= |\sum_{j=1}^{\infty} (L)\int_{I_j} \{f(x_j)-f(x)\}dx| \le \sum_{j=1}^{\infty} (L)\int_{I_j} |f(x_j)-f(x)|dx$$

$$\leq \sum_{j=1}^{\infty} (L)\int_{I_j \cap X_{k(j)}} |f(x_j)-f(x)|dx$$

$$+ \sum_{j=1}^{\infty} (L)\int_{I_j \setminus X_{k(j)}} |f(x_j)|dx + \sum_{j=1}^{\infty} (L)\int_{I_j \setminus X_{k(j)}} |f(x)|dx = R + S + T,$$

say. We show that each sum is less than $\varepsilon/3$. If $x \in I_j \cap X_{k(j)}$, $f(x)$ lies in the same range $((k(j)-1)\xi, k(j)\xi]$ as $f(x_j)$, so that they can only differ by $\pm\xi$, and

$$R \leq \sum_{j=1}^{\infty} (L)\int_{I_j \cap X_{k(j)}} \xi\, dx \leq \xi \sum_{j=1}^{\infty} m(I_j) = \xi m(M) < \varepsilon/3.$$

In S we collect those terms (if any) for which $k(j)$ has a given value k. Then

$$|f(x_j)| \leq (|k|+1)\xi,\ S = \sum_{k=-\infty}^{\infty} \sum_{k(j)=k} (L)\int_{I_j \setminus X_k} |f(x_j)|dx \leq \sum_{k=-\infty}^{\infty} \sum_{k(j)=k} (|k|+1)\xi m(I_j$$

$$\setminus X_k) \leq \sum_{k=-\infty}^{\infty} (|k|+1)\xi m(G_k \setminus X_k) < \varepsilon/3.$$

Finally, for Y the disjoint union of sets $I_j \setminus X_{k(j)}$,

$$m(Y) = \sum_{k=-\infty}^{\infty} \sum_{k(j)=k} m(I_j \setminus X_k) \leq \sum_{k=-\infty}^{\infty} m(G_k \setminus X_k) < \eta.$$

The proof is complete by definition of η, since

$$T = (L)\int_Y |f(x)|\, dx < \varepsilon/3.$$

Thus the generalized Riemann integral includes the Lebesgue integral, by taking in (3.21) a finite number of disjoint intervals forming a partition of an interval, together with suitable points x_j, forming a division of the interval. However, the generalized Riemann integral is wider as it also includes the Newton integral, and, for example, $|F'(x)|$ from (3.6) does not have a generalized Riemann integral, since

$$(2/|x|)\ |\cos(1/x^2)|$$

is not generaliz ed Riemann integrable, as can be proved. Actually it can be proved that the generalized Riemann integral is equivalent to the Perron and special Denjoy integrals when integrating $f(x)(v-u)$, and it can integrate interval-point functions.

A limitation of the generalized Riemann integral that is shared by the Riemann-Stieltjes integral and the integral by refinement of partitions, is brought out by the following theorem.

Theorem 3.5: *For* X *a set everywhere dense in* $[a,b]$ *if* f, g, g_1 *are point functions in* $[a,b]$ *with* $g \equiv g_1$ *in* X, *and if the generalized Riemann integrals in* (3.23) *exist, then*

$$(3.23)\quad \int_{[a,b)} f dg - \int_{[a,b)} f dg_1 = f(b)(g(b)-g_1(b)) - f(a)(g(a)-g_1(a)).$$

If, for two different constants G_1, G_2, $g = G_1$ *in a set everywhere dense in* $[a,b]$ *and* $g = G_2$ *in another set everywhere dense in* $[a,b]$, *the only function* f *that is generalized Riemann integrable relative to* g *in* $[a,b]$, *is a constant function.*

Clearly, if we wish to integrate more than constant functions relative to g, then g cannot oscillate too much, or else we would have to use a convergence-factor integral such as the Çesaro - Perron integral of J.C. Burkill, modified to be of Stieltjes form.

Proof: In Theorem 3.1 let the points y_j (which lie in certain intervals) be taken from X, for $j = 1,\ldots,m-1$. Then for such divisions the sums of

$$f(x_j)(g(x_j)-g_1(x_j)-g(y_{j-1})+g_1(y_{j-1})) + f(x_j)(g(y_j)-g_1(y_j)-g(x_j)+g_1(x_j))$$

$$= f(x_j)(g(y_j)-g_1(y_j)-g(y_{j-1})+g_1(y_{j-1})) = 0.$$

The only terms that do not cancel are the ones for a and b, and the generalized Riemann integral of f relative to $g-g_1$ is equal to the right of (3.23). If for the two different constants G_1 and G_2 we can replace g_1 by G_1 and by G_2, the two given values of the same integral must be the same, which gives $f(b) = f(a)$. It can be proved (see Theorem 5.1) that the integral also exists for all $[u,v] \subseteq [a,b]$, so that $f(u) = f(v)$.

Hence f has to be constant if the generalized Riemann integral exists.

Ex. 3.1. Let $f(x) = 1-g(x) = \begin{cases} 1 & (x \leq 0) \\ 0 & (x > 0) \end{cases}$. Using (3.20) for w = 0, prove that the following integrals exist.

$$\int_{[-1,1)} fdg = 1, \quad \int_{[-1,1)} gdf = 0.$$

As f and g have discontinuities at x = 0 on the same side, the refinement integral over [-1,1] does not exist. Similarly the generalized Riemann integral of f relative to g exists when f and g are both of bounded variation, though the refinement integral does not always exist.

Ex. 3.2. Let f(0) = 1, f(x) = 0(x > 0), and for an interval-point function h(I,x) in [0,1] let the generalized Riemann integral of fh over [0,1) exist with value K. Then

$$\lim_{u \to 0+} h[0,u);0) = K.$$

(Given $\varepsilon > 0$, there is a positive function δ on [0,1] such that every δ-fine division (E) of [0,1) satisfies

$$|(E) \Sigma fh - K| < \varepsilon.$$

Let $0 < u < \delta(0)$. Then there is a δ-fine division of [0,1) that uses ([0,u),0) and so

$$|h[0,u);0) - K| < \varepsilon.$$

Hence the result.)

More examples are given at the end of Section 4.

CHAPTER 2

SIMPLE PROPERTIES OF THE GENERALIZED RIEMANN INTEGRAL IN FINITE DIMENSIONAL EUCLIDEAN SPACE

4 Integration Over A Fixed Elementary Set

We begin with mutually perpendicular co-ordinate axes $Ox_1,\ldots,Ox_n$ in n-dimensional Euclidean space with the Pythagorean distance from $x = (x_1,\ldots,x_n)$ to $y = (y_1,\ldots,y_n)$,

$$\|x - y\| \equiv \surd\{\sum_{j=1}^{n} (x_j - y_j)^2\}.$$

$a = (a_1,\ldots,a_n)$ and $b = (b_1,\ldots,b_n)$ with $a_j < b_j (j = 1,\ldots,n)$, define a *brick* I, written $[a,b)$, which is the set of all points $x = (x_1,\ldots,x_n)$ with $a_j \leq x_j < b_j (j = 1,\ldots,n)$. Chapter 1, Section 3, takes $n = 1$ and bricks $[a,b)$ with $a < b$, b not being attained by points in the brick in order that for $a < b < c$, $[a,b)$ and $[b,c)$ are disjoint with union another brick $[a,c)$. Similarly we do not let x_j attain $b_j (j = 1,\ldots,n)$ when $n > 1$, while x remains in the brick. The *closure* I^c of I is the set of all x satisfying $a_j \leq x_j \leq b_j$ $(j = 1,\ldots,n)$, written $[a,b]$, and the sets of all $x \in I^c$ with $x_j = a_j$, and those with $x_j = b_j$, for a single j in $1,\ldots,n$, are called *faces* of I. The 2^n special points x with $x_j = a_j$ or $x_j = b_j$ for each $j = 1,\ldots,n$, are the *vertices of* I. If y_j has the other choice of a_j or b_j after x_j is chosen, then $y = (y_1,\ldots,y_n)$ is called the *opposite vertex to* x, and $\|x - y\|$ is the *diameter of* I, diam (I).

A finite union E of mutually disjoint bricks I is called an *elementary set,* and its *closure* E^c is the union of the corresponding I^c. *An open sphere* $S(x,r)$, *centre* x, *radius* $r > 0$, is the set of all y with $\|x - y\| < r$. Let a positive function $\delta(x)$ be defined for all $x \in E^c$. Then a brick-point pair (I,x) is *δ-fine* if $I \subseteq S(x,\delta(x))$ with x a vertex of I. *A partition* P of E is a finite collection of mutually disjoint bricks I with union E. *A division $\mathcal{D}$* of E is a finite collection of brick-point pairs (I,x), the bricks I being mutually disjoint with union E, and so forming a partition of E that, we say, *comes from $\mathcal{D}$.* Further, *$\mathcal{D}$ is δ-fine* if each $(I,x) \in \mathcal{D}$ is *δ-fine.*

Let $h(I,x)$ be a finite-valued function of brick-point pairs (I,x) with $I \subseteq E$ and $x \in E^c$. We then say that h *is defined for* E^c. For example, $h(I,x) = f(x)v(I)$ where, for $I = [a,b)$,

$$v(I) = \prod_{j=1}^{n} (b_j - a_j),$$

the *volume of* I. Then the *generalized Riemann integral of* h *over* E is a number H with the following property. Given $\varepsilon > 0$, there is a positive function δ on E^c such that

$$(4.1) \qquad |(\mathcal{D}) \Sigma h(I,x) - H| < \varepsilon$$

for all δ-fine divisions $\mathcal{D}$ of E. Here, $(\mathcal{D}) \Sigma h(I,x)$ is the sum of $h(I,x)$ for all $(I,x) \in \mathcal{D}$. This definition (4.1) can apply to any space K of values of h that has an addition (or multiplication using Π instead of Σ) together with a metric or topology to replace the modulus in (4.1). We use the notation

$$H = \int_E dh, \text{ or } H = \int_E f dk \quad \text{if} \quad h(I,x) = f(x)k(I,x) \text{ or } f(x)k(I).$$

If h and $|h|$ are (generalized Riemann) integrable over E, we say that H *is an absolute integral.* If h is integrable but not $|h|$, we say that H *is a non-absolute integral.* We sometimes use the set $S(h,\delta;E)$ of values $(\mathcal{D}) \Sigma h$ for all δ-fine divisions $\mathcal{D}$ of E.

We need the following theorem to make the definitions viable.

Theorem 4.1: *Given a positive function* δ *on* E^c, *there is a* δ*-fine division of* E.

Proof: Suppose the theorem false. Then as E is a finite union of mutually disjoint bricks I, if a δ-fine division of each such I exists, the union of the separate divisions is a δ-fine division of E. Thus the theorem is false for a brick I, *i.e.* there is no δ-fine division of I. We now use repeated bisection, we bisect I in the x_j direction by the hyperplane $x_j = \frac{1}{2}(a_j + b_j)$, with x_k arbitrary for all $k \neq j$, and repeat this for $j = 1,\dots,n$, obtaining 2^n smaller bricks J. By the previous argument from E to I, one of the J has no δ-fine division. Repeating this construction, we have a sequence

$$I = I^1 \supseteq I^2 \supseteq I^3 \supseteq \dots$$

of bricks with no δ-fine division. For (x^j) the sequence of centres of the bricks, $x^k \in I^j (k \geq j)$, $\|x^k - x^j\| \leq \operatorname{diam}(I^j) = 2^{1-j} \operatorname{diam}(I^1) \to 0$ $(j \to \infty)$, and (x^j) is a fundamental and so convergent sequence, say with limit x. $\delta(x) > 0$ exists as $x^k \in I^j \subseteq I \subseteq E$ $(k \geq j)$, $x \in E^c$; and $x \in I^{jc}$, $\operatorname{diam}(I^j) \to 0$, $I^j \subseteq S(x,\delta(x))$, eventually. For $m = 1,2,\dots,n$ the hyperplane loci of y given by

$$(y_m = x_m,\ y_k \text{ arbitrary for } k \neq m)$$

cut up I^j into at most 2^n smaller bricks J each with vertex x such that (J,x) is δ-fine. Thus there is a δ-fine division of I^j, contrary to its definition. This contradiction proves the theorem. When n = 1, Theorem 3.1 gives another proof.

Theorem 4.2: *The integral is uniquely defined.*

Proof: Let h(I,x) be integrable to H_1 and to H_2, both over the same E. Then there are positive functions δ_k on E^c and depending on $\varepsilon > 0$, such that for δ_k-fine divisions $\mathcal{D}_k$,

$$(4.2) \qquad |(\mathcal{D}_k)\Sigma h(I,x) - H_k| < \varepsilon \ (k = 1,2).$$

Then $\delta(x) \equiv \min(\delta_1(x), \delta_2(x)) > 0$ $(x \in E^c)$ and by Theorem 4.1, there is a δ-fine division $\mathcal{D}$ of E. As $\mathcal{D}$ is also δ_1-fine and δ_2-fine, the spheres being the same or larger, (4.2) is true for $\mathcal{D} = \mathcal{D}_1 = \mathcal{D}_2$. Hence

$$|H_1 - H_2| = |\{(\mathcal{D})\Sigma h - H_2\} - \{(\mathcal{D})\Sigma h - H_1\}| < 2\varepsilon,$$

while H_1, H_2 are independent of $\varepsilon > 0$. Hence $H_1 = H_2$, giving the uniqueness.

Theorem 4.3: *Let the function* h *of brick-point pairs, defined for* E, *with real or complex values, have the property that given* $\varepsilon > 0$, *there is a positive function* δ *on* E^c *such that for each pair* $\mathcal{D}$, $\mathcal{D}'$ *of* δ*-fine divisions of* E,

(4.3) $\quad |(\mathcal{D})\,\Sigma h(I,x) - (\mathcal{D}')\,\Sigma h(I,x)| < \varepsilon.$

(We then say that $(\mathcal{D})\,\Sigma h$ *is fundamental* (E).). *Then* h *is integrable over* E.

Proof: Let δ_j be a positive function on E^c such that $\varepsilon = 1/j$ in (4.3). As

$$\min(\delta_1(x),\delta_2(x),\ldots,\delta_j(x)) > 0,$$

it can replace $\delta_j(x)$, and so we can suppose that at each $x \in E^c$,

$$\delta_1(x) \geq \delta_2(x) \geq \ldots \geq \delta_j(x) \geq \ldots$$

Then each δ_k-fine division of E is δ_j-fine for all $j < k$. For each j choose a δ_j-fine division $\mathcal{D}_j$ of E, and so a sum

$$s_j = (\mathcal{D}_j)\,\Sigma h(I,x). \quad |s_j - s_k| < 1/j \ (k > j)$$

from (4.3) as $\mathcal{D}_k$ is δ_j-fine ($k \geq j$). Hence (s_j) is a fundamental and so convergent sequence, say with limit s. By (4.3) again, with δ_j-fine divisions $\mathcal{D}$ of E,

$$|(\mathcal{D})\,\Sigma h - s_k| < 1/j \ (\text{all } k \geq j), \ |(\mathcal{D})\,\Sigma h - s| \leq 1/j,$$

and we can take $j = 1/2, \ldots$ Hence by definition the integral of h exists over E with value s.

A similar result holds if the space of values of h is a more general complete space with addition and a suitable topology.

Theorem 4.4: *If* h_1 *and* h_2 *are two brick-point functions defined for* E, *that are respectively integrable to* H_1, H_2 *over* E, *and if* c *is a constant, then*

(4.4) $\quad h_1 + h_2$ *and* ch_1 *are respectively integrable to* $H_1 + H_2$ *and* cH_1.

(4.5) *If also* h_1, h_2 *are real-valued with* $h_1 \geq h_2$ *for all relevant* (I,x), *then* $H_1 \geq H_2$.

Proof: Given $\varepsilon > 0$, there is a positive function δ_j on E^c such that

(4.6) $$|(\mathcal{D}_j)\,\Sigma h_j(I,x) - H_j| < \tfrac{1}{2}\varepsilon$$

for all δ_j-fine divisions $\mathcal{D}_j$ of E $(j = 1,2)$. Taking $\delta(x) = \min(\delta_1(x), \delta_2(x)) > 0$, a δ-fine division $\mathcal{D}$ exists (Theorem 4.1), and $\mathcal{D}$ is δ_j-fine $(j = 1,2)$. Hence

$$|(\mathcal{D})\,\Sigma(h_1+h_2) - (H_1+H_2)| = |\{(\mathcal{D})\,\Sigma h_1 - H_1\} + \{(\mathcal{D})\,\Sigma h_2 - H_2\}| < \varepsilon,$$

$$|(\mathcal{D})\,\Sigma ch_1 - cH_1| = |c|\,|(\mathcal{D})\,\Sigma h_1 - H_1| \leq \tfrac{1}{2}|c|\,\varepsilon \to 0$$

with ε. Hence the results in (4.4). For the third result,

$$H_1 + \tfrac{1}{2}\varepsilon > (\mathcal{D})\ h_1 \geq (\mathcal{D})\,\Sigma h_2 > H_2 - \tfrac{1}{2}\varepsilon,\ H_1 - H_2 > -\varepsilon.$$

As $H_1 - H_2$ does not depend on $\varepsilon > 0$, it follows that $H_1 - H_2 \geq 0$.

In (4.4), h_1 and h_2 can be complex-valued with c complex, and for such functions there is a partial converse to the first part of (4.4).

Theorem 4.5: *Let* k_1, k_2 *be real-valued brick-point functions defined for an elementary set* E, *with* $k = k_1 + ik_2$. *If the complex-valued* k *is integrable over* E *to* K *then* k_j *is integrable over* E *to* $K_j (j = 1,2)$ *where* K_1, K_2 *are the real and imaginary parts of* K.

Proof: First, if a,b are real then

$$|a + ib| = \sqrt{(a^2 + b^2)} \geq \max(|a|, |b|).$$

Hence, given $\varepsilon > 0$, there is a positive function δ on E^c such that for all δ-fine divisions $\mathcal{D}$ of E,

$$\varepsilon > |(\mathcal{D})\Sigma(k_1+ik_2)-(K_1+iK_2)| = |\{(\mathcal{D})\Sigma k_1-K_1\} + i\{(\mathcal{D})\Sigma k_2-K_2\}| \geq |(\mathcal{D})\Sigma k_j-K_j|$$

$(j = 1,2)$. Being true for each $\varepsilon > 0$, by definition K_j is the integral of $k_j (j = 1,2)$.

We now have a geometrical result vital to some parts of the theory.

Theorem 4.6: (4.7) *For two disjoint elementary sets* E_1, E_2 *there is a positive function* δ_R *on* $E_1^c \cup E_2^c$ *such that if* (I,x) *is* δ_R*-fine then* $I \subseteq E_1$ *or* $I \subseteq E_2$.

(4.8) *Let bricks* $I_1,\ldots,I_m$ *form a partition* P *of an elementary set* E. *Then there is a positive function* δ_R *on* E^c *such that every* δ_R*-fine division* $\mathcal{D}$ *of* E *is a refinement of* P, *i.e. if* $(I,x) \in \mathcal{D}$ *then* $I \subseteq I_j$ *for some* j *in* $1,\ldots,m$. *We write* $\mathcal{D} \leq P$.

Proof: Let E_j^o be the interior of E_j and E_j^b the boundary or frontier $E_j^c \setminus E_j^o$ of E_j. As E_j is a finite union of bricks, the geometry shows that E_j^b is a finite union of parts of faces of the bricks, the parts being $(n-1)$-dimensional bricks. If $x \in E_1^o \cup E_2^o$ let $2\delta_R(x)$ be the distance from x to $E_1^b \cup E_2^b$. If $x \in E_1^b \cup E_2^b$ let $2\delta_R(x)$ be the minimum distance from x to the parts of faces in $E_1^b \cup E_2^b$ on which x does not lie. Then $\delta_R(x) > 0$. If (I,x) is δ_R-fine and I_c has no points in common with $E_1^b \cup E_2^b$ then $x \in E_1^o$, $I \subseteq E_1$, or $x \in E_2^o$, $I \subseteq E_2$). If (I,x) is δ_R-fine and $I^c \cap (E_1^b \cup E_2^b)$ is not empty, then by construction $x \in E_1^b \cup E_2^b$ and I^c does not intersect any part of a face in $E_1^b \cup E_2^b$ on which x does not lie. If x lies on face F, then as x is a vertex of I, no interior point of I lies on F. As $I \subseteq E_1 \cup E_2$ then I cannot have interior points in E_1 and other interior points in E_2, or else a face would have interior points of I. Hence either $I \subseteq E_1$ or $I \subseteq E_2$, proving (4.7).

To prove (4.8) let δ_j be the δ_R of (4.7) for which $E_1 = I_j$ and $E_2 = E\setminus I_j$ Then $\min(\delta_1,\ldots,\delta_m) > 0$ is the δ_R for (4.8).

<u>Corollary</u>: *Every refinement or σ-integral in $\mathbb{R}^n$ using bricks, is a generalized Riemann integral.*

In many parts of analysis, monotonicity plus boundedness often leads to convergence. Here is an example. First, a real-valued function $h(I)$ of the bricks $I \subseteq E$, is finitely sub-additive if, for each $I \subseteq E$ and each partition P of I,

$$h(I) \leq (P)\Sigma h.$$

If the inequality is always reversed, h is finitely superadditive.

<u>Theorem 4.7</u>: *If the real-valued* h *is a finitely subadditive brick function in* E *and if, for some positive function* δ *on* E^c *and all* δ*-fine divisions* $\mathcal{D}$ *of* E, *the set of sums*

$$s = (\mathcal{D})\Sigma h(I)$$

is bounded above with supremum H, *then* h *is integrable over* E *to the value* H. *If, on the other hand, the set of sums* s *is unbounded, then for each integer* m *there is a positive function* δ_m *on* E^c *such that* $s \geq m$ *for all sums* s *of* h *over* δ_m*-fine divisions of* E.

<u>Proof</u>: Given a division $\mathcal{D}$ over E, and so a partition P and a sum s, by Theorem 4.6 (4.8) there is a positive function δ_R on E^c such that all δ_R-fine divisions $\mathcal{D}'$ of E satisfy $\mathcal{D}' \leq P$. If sums s are bounded above by a supremum H then, given $\varepsilon > 0$, there is an s lying in $H-\varepsilon < s \leq H$. As h is finitely sub-additive,

$$H-\varepsilon < (P)\Sigma h(I) \leq (\mathcal{D}')\Sigma h(I) \leq H,$$

and by definition h is integrable to H. On the other hand, if sums are unbounded above, for each integer m there is an $s \geq m$, from a partition P of E. Thus

$$m \leq s = (P)\Sigma h \leq (\mathcal{D}')\Sigma h(I)$$

for all δ_R-fine divisions $\mathcal{D}'$, finishing the proof.

The only difference between Riemann and generalized Riemann integration lies in the fact that the first uses a constant mesh over the whole of the elementary set E on which the integral is defined, whereas the second uses a variable mesh controlled by a positive function δ. It is therefore useful to look at a few examples of such positive functions. Two are given in Section 3, and here we generalize them to R^n.

Ex. 4.1. For $w \in E^c$ suppose that $\delta(x) < |x-w|$ when $x \neq w$. Then a δ-fine (I,x) can only have $w \in I^c$ when $x = w$. Thus we force w to be a vertex of at least one brick from each δ-fine division of E. We can apply this construction to a finite number $w_1, w_2, \ldots, w_j$ of points of E^c to obtain a function $\delta_j > 0$, or to a sequence (w_j), that could be dense in E^c, by using the corresponding sequence (δ_j).

Ex. 4.2. For G an open set in R^n and each $x \in G$, we can arrange that the sphere

$$S(x,\delta(x)) \subseteq G.$$

Ex. 4.3. If $\delta_j (j = 1,\ldots,m)$ are positive functions on E^c, so is $\min(\delta_1,\ldots,\delta_m)$.

The same cannot be said of an infinite sequence of positive functions unless the sequence has special properties. For if $\delta_j(x) = 1/j (x \in E^c)$, the minimum is 0.

Ex. 4.4. If a positive function $\delta_y(x) (x \in E^c)$ is given for each $y \in E^c$, $\delta(x) \equiv \delta_x(x)$ is also a positive function. This is like a Cantor diagonal process.

Ex. 4.5. In one dimension let $P_m(\mathcal{D}_m)$ be the families of all $[u,v)$ (respectively, all $([u,v),u)$ and all $([u,v),v)$) that satisfy

$$0 < u < v \leq 1,\ u(m+1)/m < v \leq um/(m-1)\ (m \geq 2)\ .$$

If $0 < \delta(x) \leq x/(m+1)$ $(x > 0)$ then no member of $\mathcal{D}_m$ can be δ-fine. For if $v-u < u/(m+1)$ then $v < u(m+2)/(m+1) < u(m+1)/m$. If $v-u < v/(m+1)$ then $vm/(m+1) < u$.

Ex. 4.6. Let P and $\mathcal{D}$ be the respective unions of the P_m and $\mathcal{D}_m$, for $m = 2,3,\ldots$, in *Ex.* 4.5. Then P is the set of $[u,v)$ with $u < v \leq 2u$, $0 < u < v \leq 1$, and $\mathcal{D}$ includes the end-points u,v as well.

Ex. 4.7. If the integrable $h_m(I,x) \geq 0$ for all δ-fine (I,x) in E^c and some $\delta(x) > 0$, with

$$h(I,x) = \sum_{m=1}^{\infty} h_m(I,x),$$

finite and integrable, then

$$\int_E dh \geq \sum_{m-1}^{\infty} \int_E dh_m. \tag{4.9}$$

For in Theorem 4.4 (4.5) and each integer m, we use

$$h \geq \sum_{j=1}^{m} h_j.$$

Ex. 4.8. There may be inequality in (4.9). For in *Ex.* 4.5 let

$$h_m([u,v),x) = \begin{cases} v-u \ ([u,v) \in P_m,\ m > 1) \\ 0 \ (\text{otherwise}) \end{cases} \qquad h([u,v),x) = \begin{cases} v-u \ (u < v \leq 2u) \\ 0 \ (\text{otherwise}) \end{cases}.$$

$$\int_{[0,1)} dh_m = 0, \quad \int_{[0,1)} dh = 1.$$

Ex. 4.9. In *Ex.* 4.8, if

$$k = \sum_{m=1}^{\infty} h_{2m},$$

then k is not integrable in any interval of $[0,1)$.

For, given a positive function δ on $[0,1]$, let S be the non-empty set of all x in $[0,1]$ such that either $x = 1$ or the supremum of sums of the k over δ-fine divisions of $[x,1)$ is $1-x$. If s is the infimum of S then $0 \leq s \leq 1$ and $s \in S$. If $s > 0$, there is an integer m such that

$$(4.10) \qquad ([(2m-1)s/(2m),s)s) \in \mathcal{D}_{2m}$$

and is δ-fine. Thus $(2m-1)s/(2m) \in S$, contrary to the definition of s. Hence $s = 0$ and the supremum, of sums of k over δ-fine divisions of $[x,y) \subseteq [0,1)$, is $y-x$. Similarly, replacing $2m$ by $2m+1$ in (4.10), we prove that the infimum of sums of the k over δ-fine divisions of $[x,y)$, is 0. Thus the generalized Riemann integral cannot exist over $[x,y)$.

Ex. 4.10. In R^n let f be a point function and $h \geq 0$ a brick-point function, both on E^c. If $X \subseteq R^n$, if $f = 0$ in $\sim X$ (so that $f = f\chi(X;\cdot)$) and if fh and $\chi(X;\cdot)h$ are integrable over E then

$$m \int_E \chi(X;.)dh \leq \int_E fdh \leq M \int_E \chi(X;.)dh$$

where $m = \inf f$, $M = \sup f$, on X. For on X,

$$mh \leq fh \leq Mh.$$

(Exceptionally, we could be dealing with infinite values, of m or M or both. The inequalities only have value when at least one of m,M is finite, and then it can be proved that the integral of fh is equal to a Lebesgue integral over $X \cap E$ of f with respect to the measure given for varying bricks $I \subseteq E$ by

$$\int_I \chi(X;.)dh.$$

Ex. 4.11. If in *Ex.* 4.10 f is continuous on $X = E^c$, where E is a brick, there is a point $\xi \in E^c$ with

$$\int_E fdh = f(\xi) \int_E dh.$$

For f takes all values from m to M, and in particular the value

$$\int_E fdh \bigg/ \int_E dh.$$

5 Integration And Variation Over More Than One Elementary Set

A real or complex valued function h of elementary sets (possibly in a fixed elementary set) *is finitely additive* if for each pair E_1, E_2 of disjoint elementary sets, so that $E_1 \cup E_2$ is also an elementary set, we have

$$(5.1) \qquad h(E_1) + h(E_2) = h(E_1 \cup E_2).$$

Such functions are most useful, in that a sum of them over partitions of an elementary set E, is equal to h(E). If h is real-valued and if instead of (5.1) we always have

$$(5.2) \qquad h(E_1) + h(E_2) \le h(E_1 \cup E_2),$$

we say that h *is finitely superadditive.* On the other hand, if the inequality always goes the other way,

$$(5.3) \qquad h(E_1) + h(E_2) \ge h(E_1 \cup E_2),$$

we say that h *is finitely subadditive.* In the case (5.2), refinement of partitions implies that the sums over partitions fall, while in case (5.3) the sums rise.

We say that E_1 *is a partial set of* E if both E_1 and E are elementary sets and $E_1 \subseteq E$. If also $E_1 \ne E$ we say that E_1 *is a proper partial set of* E.

Theorem 5.1: *If a function* h(I,x) *of brick-point pairs is integrable over an elementary set* E *to* H(E), *then* h *is integrable over every partial set* E_1 *of* E, *say to* $H(E_1)$, *and* H *is finitely additive over the partial sets* E_1.

Proof: Let E_1 be a proper partial set of E. Then by the geometry, $E \setminus E_1$ is also a partial set. Given $\varepsilon > 0$, if the positive δ, defined on E^c, is such that

$$(5.4) \qquad |(\mathcal{D})\, \Sigma h(I,x) - H(E)| < \varepsilon$$

for all δ-fine divisions $\mathcal{D}$ of E, then by Theorem 4.1 there are δ-fine divisions

$\mathcal{D}_1$ of E_1 and $\mathcal{D}_2$ of $E \setminus E_1$, and $\mathcal{D}_1 \cup \mathcal{D}_2$ is a δ-fine division of E. Also using a second δ-fine division $\mathcal{D}_3$ of E_1 with (5.4), if

$$s_j = (\mathcal{D}_j)\Sigma h \ (j=1,2,3), \ |s_1-s_3| = |(s_1+s_2-H(E))-(s_3+s_2-H(E))| < 2\varepsilon, \ (\mathcal{D}_1)\Sigma h$$

is fundamental (E_1), and by Theorem 4.3 the integral over E_1 exists, say as $H(E_1)$. Letting $s_3 \to H(E_1)$, we have

$$(5.5) \qquad |s_1 - H(E_1)| \leq 2\varepsilon.$$

With the same δ, ε, this is true for every partial set E_1 of E. If E_4, E_5 are disjoint partial sets of E, so that $E_4 \cup E_5$ is a partial set, and if s_j is a sum over a δ-fine division of $E_j (j = 4,5)$ then s_4+s_5 is a sum over a δ-fine division of $E_4 \cup E_5$, and (5.5) is true for each of the three partial sets. Hence

$$|H(E_4 \cup E_5)-H(E_4)-H(E_5)| = |\{H(E_4 \cup E_5)-(s_4+s_5)\} + \{s_4-H(E_4)\} + \{s_5-H(E_5)\}| \leq 6\varepsilon$$

But the values of the integrals are independent of $\varepsilon > 0$, so that

$$H(E_4 \cup E_5) - H(E_4) - H(E_5) = 0$$

and H is finitely additive over partial sets of E.

Clearly this theorem is vital, and it is directed inwards from E to its partial sets. We now go outwards, to unions of elementary sets.

Theorem 5.2: *If* E_1, E_2 *are disjoint elementary sets with a brick-point function* h *integrable over* E_1 *and over* E_2, *then* h *is integrable over* $E_1 \cup E_2$.

Proof: Given $\varepsilon > 0$, there is a positive function δ_j on E_j^c, such that for all δ_j-fine divisions $\mathcal{D}_j$ of E_j,

$$|(\mathcal{D}_j)\Sigma h - H(E_j)| < \tfrac{1}{2}\varepsilon \ (j = 1,2).$$

Let δ_R be the positive function of Theorem 4.6 (4.7) and put $\delta(x)$ equal to $\min(\delta_R(x), \delta_1(x))$ in $E_1^c \setminus E_2^c$, $\min(\delta_R(x), \delta_2(x))$ in $E_2^c \setminus E_1^c$, and

$\min(\delta_R(x), \delta_1(x), \delta_2(x))$ in $E_1{}^c \cap E_2{}^c$, $= E_1{}^b \cap E_2{}^b$ as the interiors are disjoint. This defines $\delta > 0$ in $E_1{}^c \cup E_2{}^c$, and if $\mathcal{D}$ is a δ-fine division of $E_1 \cup E_2$, $\mathcal{D}$ splits up into a δ_j-fine division $\mathcal{D}_j$ of E_j $(j = 1,2)$. Thus

$$|(\mathcal{D})\Sigma h - H(E_1) - H(E_2)| = |(\mathcal{D}_1)\Sigma h - H(E_1) + (\mathcal{D}_2)\Sigma h - H(E_2)| < \varepsilon.$$

proving the result.

Theorem 5.3: *Given* $\varepsilon > 0$, *let* (5.4) *hold for all* δ*-fine divisions* $\mathcal{D}$ *of* E, *where* h *is real or complex valued. Then*

$$(5.6) \qquad (\mathcal{D})\ \Sigma|h(I,x) - H(I)| \leq 8\varepsilon$$

for all δ*-fine divisions* $\mathcal{D}$ *of* E. *Conversely, let* $h(I,x)$ *and* $H(E_1)$ *be given with* H *finitely additive over partial sets* E_1 *of* E. *If, given* $\varepsilon > 0$, (5.6) *is true for some positive function* δ *on* E^c, *then* h *is integrable to* H *on all partial sets of* E.

Proof: Theorem 5.1 gives (5.5) and the finite additivity of the integral H. Thus if $\mathcal{D}_1$ is a δ-fine division of a partial set of E then

$$|(\mathcal{D}_1)\ \Sigma\{h(I,x) - H(I)\}| \leq 2\varepsilon.$$

Let $\mathcal{D}$ be a δ-fine division of E and let $\mathcal{D}_1$ be that part of $\mathcal{D}$ with $\text{real}(h-H) \geq 0$. Then $(\mathcal{D}_1)\ \Sigma\,|\text{real}(h-H)| = \text{real}\ (\mathcal{D}_1)\ \Sigma\,(h-H) \leq |(\mathcal{D}_1)\ \Sigma\,(h-H)| \leq 2\varepsilon$,

$$(\mathcal{D}\setminus\mathcal{D}_1)\ \Sigma\,|\text{real}(h-H)| = (\mathcal{D}\setminus\mathcal{D}_1)\ \Sigma\,\text{real}(H-h) = \text{real}\ (\mathcal{D}\setminus\mathcal{D}_1)\ \Sigma\,(H-h)$$

$$\leq |(\mathcal{D}\setminus\mathcal{D}_1)\ \Sigma\ (h-H)| \leq 2\varepsilon, \quad (\mathcal{D})\ \Sigma\,|\text{real}(h-H)| \leq 4\varepsilon.$$

Similarly

$$(\mathcal{D})\ \Sigma\,|\text{imag}(h-H)| \leq 4\varepsilon, \text{ and so } (\mathcal{D})\ \Sigma\,|h-H| \leq 8\varepsilon.$$

Conversely, if $E_1 \neq E$ let $\mathcal{D}_1$, $\mathcal{D}_2$ be δ-fine divisions of E_1 and $E\setminus E_1$. Then $\mathcal{D}_1 \cup \mathcal{D}_2$ is a δ-fine division of E. By finite additivity of H, h is integrable to $H(E_1)$ over E_1 since

$$|(\mathcal{D}_1)\Sigma h - H(P)| = |(\mathcal{D}_1)\Sigma\{h(I,x)-H(I)\}| \le (\mathcal{D}_1)\Sigma|h-H| \le (\mathcal{D}_1 \cup \mathcal{D}_2)\Sigma|h-H| \le 8\varepsilon.$$

A function H of partial sets of E, is the *variational integral* of a brick-point function h in E, if H is finitely additive over partial sets of E and if, given $\varepsilon > 0$, there is a positive function δ on E^c such that (5.6) is true for all δ-fine divisions $\mathcal{D}$ of E. In this notation Theorem 5.3 says that the variational and generalized Riemann integrals are equivalent.

A second definition of the variational integral is as follows. H is the *variational integral of* h if H is finitely additive over partial sets of E and if, given $\varepsilon > 0$, there are a finitely superadditive function S of partial sets of E with $S(E) \le 8\varepsilon$, and a positive function δ on E^c, such that for all δ-fine (I,x),

$$(5.7) \qquad |h(I,x) - H(I)| \le S(I).$$

Clearly (5.7) implies (5.6), while (5.6) implies (5.7) by using

$$S(E_1) = \sup\ (\mathcal{D}')\Sigma|h-H|$$

over all δ-fine divisions $\mathcal{D}'$ of E_1. For if E_1, E_2 are disjoint, divisions $\mathcal{D}_j$ over E_j ($j = 1,2$) give the division $\mathcal{D}_1 \cup \mathcal{D}_2$ of $E_1 \cup E_2$. But not every division of $E_1 \cup E_2$ can be split up into divisions $\mathcal{D}_j$ of E_j ($j = 1,2$) as the δ could sometimes be greater than a δ_R.

Theorem 5.3 also leads to the variation. Given a positive function δ on E^c let

$$V(h;\delta;E) = \sup\ (\mathcal{D})\Sigma|h(I,x)|,$$

for the supremum over all δ-fine divisions $\mathcal{D}$ of E. *The variation of* h *in* E *is*

$$V(h;E) = \inf_{\delta>0} V(h;\delta;E), = \limsup_{\delta \to 0+} (\mathcal{D})\Sigma|h(I,x)|.$$

The supremum and infimum could be a conventional $+\infty$. If the variation is finite, so that $V(h;\delta;E)$ is finite for some positive function δ on E^c, we say that h *is of bounded variation* (VB*) *in* E. If the variation is 0, we say that h *is of variation zero in* E. Two brick-point functions h,k for the same E^c are *variationally equivalent* there if h-k is of variation zero in E.

If X is a set in n-dimensions with indicator $\chi(X;x)$ we write, respectively, $V(h;\delta;E;X)$, $V(h;E;X)$ for $V(h.\chi(X;.);\delta;E)$, $V(h.\chi(X;.)E)$. If $V(h;E;X)$ is finite, we say that h *is of bounded variation in* X, *relative to* E.

There should be no confusion between $V(h;E;X)$ and $V(h;\delta;E)$ as the latter has a Greek letter. We say that a property is true h-*almost everywhere* (h-*a.e.*) if it is true except in a set X with $V(h;E;X) = 0$. Such a set X is said to be of h-*variation zero.*

Much interaction between integrals and variations occurs, as the following shows.

Theorem 5.4: (5.8) *If the brick-point function* h *for* E^c *is integrable to* $H(E_1)$ *over each partial set* E_1 , *then* h-H *is of variation zero and* h *is variationally equivalent to* H. *Also*

(5.9) $$V(h;E;X) = V(H;E;X)$$

for each set X *in* n *dimensions, even if one side is* $+\infty$.

(5.10) *If* $V(h;E) = 0$ *and if* E_1 *is a partial set of* E, *then* $V(h,E_1) = 0$ *and there exists*

(5.11) $$\int_{E_1} dh = 0.$$

(5.12) *Conversely, if* (5.11) *is true for every partial set* E_1 *of* E *then* $V(h;E) = 0$.

(5.13) *If* $|h|$ *is integrable to* K *over* E *then* $V(h;E)$ *is finite and equal to* K.

(5.14) *Conversely, if in* (5.8) $V(h;E)$ *is finite then* $|h|$ *is integrable, and*

(5.15) $$|H(E_1)| = \left|\int_{E_1} dh\right| \leq \int_{E_1} d|h| = V(h;E_1), = V^*(E_1) \ (say).$$

(5.16) $$V(h;E;X) = V(H;E;X) = V(V^*;E;X).$$

Proof: (5.8) is just a rewriting of Theorem 5.3 (5.6) in two ways using the variation. For (5.9) we multiply by $\chi(X;.)$ the inequalities

(5.17) $$|h| \leq |h-H| + |H| , \ |H| \leq |H-h| + |h|$$

and then sum over a division of E, showing that if one side of (5.9) is finite, so is the other, and then that the two are equal. For (5.10) and (5.11) we use a δ-fine division $\mathcal{D}_1$ of E_1, and when $E_1 \neq E$, a δ-fine division $\mathcal{D}_2$ of $E \setminus E_1$ with

(5.18) $$|(\mathcal{D}_1)\Sigma h| \leq (\mathcal{D}_1)\Sigma |h| \leq (\mathcal{D}_1 \cup \mathcal{D}_2)\Sigma |h| .$$

Hence $V(hE_1) = 0$ and $H(E_1) = 0$ follow from $V(h;E) = 0$. For (5.12), $H(I) = 0$ for all $I \subseteq E$, and (5.8) gives (5.12). For (5.13), given $\varepsilon > 0$, there is a positive function δ on E^c such that if $\mathcal{D}$ is a δ-fine division of E,

$$K-\varepsilon < (\mathcal{D})\Sigma|h| < K+\varepsilon, \ K-\varepsilon < V(h;\delta;E) \leq K+\varepsilon, \ K-\varepsilon \leq V(h;E) \leq K+\varepsilon,$$

and $K = V(h;E)$, finite. In (5.14) we use (5.17) to show that $|h|$ is integrable over E if and only if $|H|$ is integrable over E, to the same value. By Theorem 5.1, H is finitely additive over partial sets of E, so that if $\mathcal{D}_1$ is a division of a partial set E_1 of E,

$$|H(E_1)| = |(\mathcal{D}_1)\Sigma H(I)| \leq (\mathcal{D}_1)\Sigma |H(I)|$$

and $|H|$ is finitely subadditive. Hence for each division $\mathcal{D}$ of E

$$(\mathcal{D})\Sigma |H(I)| \leq V(H;E), = V(h;E)$$

using (5.9) with $X = R^n$. Hence by Theorem 4.7, $|H|$ is integrable, and so $|h|$ is integrable. (5.15) follows by (5.18) and the integrability. Then in (5.16), as in (5.9),

$$V(h;E;X) = V(|h|;E;X) = V(V^*;E;X).$$

A special case of (5.13), (5.14) is well worth setting out as a separate theorem using the volume $v(I)$ of the brick I.

Theorem 5.5: *Let* $h(I,x) = f(x)v(I)$ *be integrable over an elementary set* E *with integral* $H(E_1)$ *over* E_1. *Then* $|f(x)|v(I)$ *is integrable over* E *if and only if* H *is of bounded variation over* E.

Non-absolute integrals H are not of bounded variation, and for these we need another definition. The brick-point function h on E^c *is of generalized bounded variation* (VBG*) *in a set* X *relative to* E, if X is the union of an at most countable number of sets X_j on each of which h is of bounded variation relative to E.

If the sum of the $V(h;E;X_j)$ is finite, the next theorem shows that h is of bounded variation on X relative to E and we go no further.

Theorem 5.6: *If* h *is a brick-point function on* E^c *and if* (X_j) *is a sequence of sets in* R^n *with union* X, *then*

$$V(h;E;X) \leqq \sum_{j=1}^{\infty} V(h;E;X_j).$$

Proof: If some X_j have points in common, the left-hand side is unaltered, but the right-hand side will tend to increase. Thus we can assume mutually disjoint X_j, and a finite right-hand side or else there is nothing to prove. Given $\varepsilon > 0$, let the positive function δ_j on E^c be such that

(5.19) $$V(h;\delta_j;E;X_j) < V(h;E;X_j) + \varepsilon.2^{-j} \quad (j = 1,2,\ldots).$$

We take

(5.20) $$\delta(x) = \begin{cases} \delta_j(x) & (x \in X_j,\ j = 1,2,\ldots), \\ 1 & (x \in \setminus X). \end{cases}$$

If $\mathcal{D}$ is a δ-fine division of E and if Q_j is the subset of the $(I,x) \in \mathcal{D}$ with $x \in X_j$, then as $\mathcal{D}$ is only a finite set, there is an integer m with Q_j empty for $j > m$. Hence by (5.19),

$$(\mathcal{D}) \Sigma |h(I,x)| \chi(X;x) = \sum_{j=1}^{m} (Q_j) \Sigma |h(I,x)| \le \sum_{j=1}^{m} V(h;\delta_j;E;X_j)$$

$$< \sum_{j=1}^{\infty} V(h;E;X_j) + \varepsilon, \quad V(h;\delta;E;X) \le \sum_{j=1}^{\infty} V(h;E;X_j) + \varepsilon.$$

Letting δ and ε tend to 0, we have the theorem.

In the language of Lebesgue theory, Theorem 5.6 shows that the variation is an outer measure. We use this to prove some results connected with integration by substitution. In particular, the integrals of $f(x)\,k(I,x)$ and $g(x)\,k(I,x)$ are equal if $f = g$ k-almost everywhere.

Theorem 5.7: *If* f *is a finite-valued point function on* E^c *and* k *is a brick-point function such that*

(5.21) $\quad V(k;E;X) = 0,$ *then*

(5.22) $\quad V(fk;E;X) = 0, \quad \int_E f.\chi(X;.)dk = 0.$

Conversely, if (5.22) *is true and if* X_0 *is the set where* $f \neq 0$*, then* (5.21) *is true with* X *replaced by* $X \cap X_0$.

(5.23) *If* f *is a finite-valued point function on* E^c*, if* $f(x)h(I,x)$ *is integrable over* E*, and if*

(5.24) $\quad V(h - k;E) = 0,$

then fk *is integrable over* E *and the two integrals are equal for all partial sets of* E.

(5.25) *If* $H(E_1) = \int_{E_1} g\,dh$ *exists for all partial sets* E_1 *of* E, *the existence of either of* $\int_E f\,dH$, $\int_E fg\,dh$ *implies the existence and equality of both. (Integration by substitution).*

(5.26) *If for all partial sets* E_1 *of* E *and two point functions* f,g *and a brick-point function* k *on* E^c,

$$\int_{E_1} f\,dk = \int_{E_1} g\,dk,$$

then f = g k-*almost everywhere, (*k *need not be non-negative.)*

(5.27) *If for all partial sets* E_1 *of* R, *a point function* f *and two brick-point functions* h, k *on* E^c,

$$\int_{E_1} f\,dh = \int_{E_1} f\,dk$$

then f = 0 (h-k)-*almost everywhere.*

Proof: For (5.22) let X_j be that part of X for which $j-1 \le |f| < j\ (j = 1,2,\ldots)$. Then $V(fk;\ E;X_j) \le jV(k;E;X_j) \le jV(k;E;X) = 0$.

Theorems 5.6 and 5.4 (5.11) give (5.22). For the converse we use the proof just given, replacing X by $X \cap X_0$, and k and fk by fk and $f^{-1}(fk) = k$, respectively, which can be done in X_0. For (5.23) we use (5.24) just as (5.21) is used to get (5.22). Thus

$$(5.28) \qquad \int_{E_1} f\,d(h-k) = 0$$

for all partial sets E_1 of E. Theorem 4.4 (4.4) finishes the proof. For (5.25) we use Theorem 5.4 (5.8) and substitute H and gh for h and k in (5.23). For (5.26) we use Theorem 4.4 (4.4) to obtain

$$\int_{E_1} (f-g)dk = 0$$

for all partial sets E_1 of E. Then Theorem 5.4 (5.8) gives

$$V((f-g)k;E) = V((f-g)k - \int (f-g)dk;E) = 0$$

and the result follows from (5.22) implying (5.21) with $X = E^c$ and X_0 the set where $f-g \neq 0$. (5.27) follows using Theorem 4.4 (4.4) for (5.28), and then a similar proof.

<u>Theorem 5.8</u>: *For disjoint elementary sets* E_1, E_2 *and* δ_R *the positive function of Theorem* 4.6 (4.7) *let* δ_j *be a positive function on* $E_j{}^c (j = 1,2)$, *with* δ *as in Theorem* 5.2. *Then for an arbitrary positive function* δ^* *on* $E_1{}^c \cup E_2{}^c$.

(5.29) $V(h;\delta^*;E_1) + V(h;\delta^*;E_2) \leq V(h;\delta^*;E_1 \cup E_2)$,

(5.30) $V(h;\delta;E_1) + V(h;\delta;E_2) = V(h\delta;E_1 \cup E_2)$,

(5.31) $V(h;E_1) + V(h;E_2) = V(h;E_1 \cup E_2)$.

(5.32) *If* E_1 *is a proper partial set of* E *with* $V(h;E)$ *finite,* $\varepsilon > 0$, *and* δ_3 *a positive function on* E^c *with* $V(h;\delta_3;E) < V(h;E) + \varepsilon$, *then* $V(h;\delta_3;E_1) < V(h;E_1) + \varepsilon$.

<u>Proof</u>: If s_j is a sum of $|h|$ for a δ^*-fine division of E_j $(j = 1,2)$ then $s_1 + s_2$ is a sum for a δ^*-fine division of $E_1 \cup E_2$, giving (5.29). Next, as $\delta \leq \delta_R$, a sum of $|h|$ for a δ-fine division of $E_1 \cup E_2$ gives a sum for E_1 plus a sum for E_2, and the opposite inequality to (5.29) follows to give (5.30). Given $\varepsilon > 0$, we choose δ_j on $E_j{}^c$ with

$$V(h;\delta_j;E_j) < V(h;E_j) + \tfrac{1}{2}\varepsilon,$$

Constructing δ and using (5.30),

$$V(h;E_1 \cup E_2) \leq V(h;\delta;E_1 \cup E_2) = V(h;\delta;E_1) + V(h;\delta;E_2) < V(h;E_1) + V(h;E_2) + \varepsilon,$$

(5.33) $\quad V(h;E_1 \cup E_2) \leq V(h;E_1) + V(h;E_2).$

In particular, if the right-hand side is finite, so is the left-hand side. To prove the opposite inequality there is a positive function δ_4 on $E_1{}^c \cup E_2{}^c$ such that

$$V(h;\delta_4;E_1 \cup E_2) < V(h;E_1 \cup E_2) + \varepsilon.$$

Taking $\delta_j = \delta_4$ on $E_j{}^c (j = 1,2)$ we construct δ and find from (5.30),

$$V(h;E_1) + V(h;E_2) \leq V(h;\delta;E_1) + V(h;\delta;E_2) = V(h;\delta;E_1 \cup E_2) < V(h;E_1 \cup E_2) + \varepsilon,$$

giving the opposite inequality to (5.33) and so (5.31). Using (5.29), (5.31) in (5.32),

$$V(h;\delta_3;E_1) \leq V(h;\delta_3;E) - V(h;\delta_3;E\setminus E_1) < V(h;E) + \varepsilon - V(h;E\setminus E_1) = V(h;E_1) + \varepsilon.$$

Theorem 5.8 (5.29) shows that $S(E) = V(h;\delta^*;E)$ is finitely superadditive over elementary sets E, so that there is a second definition of the variation of h, corresponding to the second definition of the variational integral, as the infimum of S(E) for all finitely superadditive functions S over partial sets of E and positive δ^* that satisfy $|h(I,x)| \leq S(I)$ for all δ^*-fine (I,x). We use this definition in the next theorem.

Theorem 5.9: *h is VBG* in X, relative to E, if and only if there are a positive function δ in E^c, a non-negative finitely superadditive function S of bricks in E with $S(E) > 0$, and a point function $k > 0$, such that for every δ-fine (I,x),*

(5.34) $\quad |h(I,x)| \leq k(x)S(I).$

Proof: If h is VBG* in X relative to E, there are sets X_j $(j = 1,2,\ldots)$ with union X, in each of which h is of bounded variation relative to E. Clearly we can take the X_j mutually disjoint. There are positive functions δ_j on E^c and finitely superadditive functions S_j such that for all δ_j-fine (I,x),

$$|h(I,x)| \le S_j(I) \quad (x \in X_j,\ j = 1,2,\dots).$$

We prove (5.34) by taking

$$\delta(x) = \delta_j(x)(x \in X_j),\ \delta(x) = 1(x \notin X),\ S(I) = \sum_{j=1}^{\infty} 2^{-j} S_j(I)/S_j(E), k(x) = 2^j S_j(E)$$

$$(x \in X_j,\ j = 1,2,\dots),\ k(x) = 1\ (x \notin X).$$

Conversely, if (5.34) is true let X_{1j} be the set where $j - 1 < k(x) \le j$. Then

$$|h(I,x)| \le jS(I)$$

and h is of bounded variation on X_{1j} for $j = 1,2,\dots$, and so is VBG* in X relative to E.

For two brick-point functions h,k in E^c and a set X of R^n, we say that h *is* k-*absolutely continuous* (k-AC*) *in* X *and* E, if for varying sets $X_0 \subseteq X$,

$$V(h;E;X_0) \to 0 \text{ when } V(k;E\cdot X_0) \to 0.$$

Further, h *is generalized* k-*absolutely continuous* (k-ACG*) *in* X *and* E, if there is a sequence (X_j) of sets with union X, such that h is k-absolutely continuous in each X_j and E.

Theorem 5.10: *If* f *is a point function and* h *a brick-point function in* E^c, *both real or complex valued, then* fh *is* h-ACG* *in* E^c *and* E, *and if* fh *is integrable to* $F(E_1)$ *over partial sets* E_1 *of* E, *then* F *is* h-ACG* *in* E^c *and* E. *If also* h *is* VB* *or* VBG* *on* X *relative to* E, *except possibly where* f = 0, *then* fh, *and* F *when* fh *is integrable, are also* VBG* *on* X *relative to* E.

Proof: For the first part, for each set X_j where $j - 1 \le f(x) < j$, fh is h-AC* in X_j and $E(j = 1,2,\dots)$, and if fh is integrable to F, Theorem 5.4 (5.9) shows that F is h-AC* relative to the same set. Next we use (5.34) in the proof of Theorem 5.9, with k multiplied by $|f|$, or by 1 if f = 0. If k does not exist at a point where f = 0, we multiply S by 1.

When $h \geq 0$ and both fh and $|f|h$ are integrable over E, we prove later that F is h-AC* in E^c and E, but the more difficult proof needs a 'limit under the integral sign' theorem, see Theorem 8.4.

There is a property of regular measures (in the Lebesgue theory) that is not always true for general measures. Here we prove it true for the variation.

Theorem 5.11: *If* (X_j) *is a monotone increasing sequence of* n-*dimensional sets with union* X, *then* $\lim_{j\to\infty} V(h;E;X_j) = V(h;E;X)$ *for every brick-point function* h *on* E^c. *More generally, for an arbitrary sequence* (X_j) *of* n-*dimensional sets,*

$$(5.35)\qquad V(h;E;\liminf_{j\to\infty} X_j) \leq \liminf_{j\to\infty} V(h;E;X_j).$$

Proof: As $X \supseteq X_j$ we have

$$(5.36)\qquad V(h;E;X_j) \leq V(h;E;X),\ \lim_{j\to\infty} V(h;E;X_j) \leq V(h;E;X).$$

Thus if the limit is $+\infty$ the result is true. If the limit is finite, we prove the opposite inequality to (5.36). Let $\delta_j(x) > 0$ on E^c satisfy

$$(5.19)\qquad V(h;\delta_j;E;X_j) < V(h;E;X_j) + \varepsilon.2^{-j}$$

for $j = 1,2,\ldots$, and define

$$\delta(x) = \delta_j(x)(x \in X_j \setminus X_{j-1},\ j = 1,2,\ldots,\ X_0 \text{ empty}),\ \delta(x) = 1\ (x \in \setminus X).$$

If $\mathcal{D}$ is a δ-fine division of E and if Q, Q_j are the subsets of $\mathcal{D}$ with the x in X and in $X_j \setminus X_{j-1}$, respectively $(j = 1,2,\ldots)$, there is a greatest integer m, depending on $\mathcal{D}$, such that Q_m is not empty. Let E_j be the union of the bricks I from the $(I,x) \in Q_j (j = 1,\ldots,m)$. Then from (5.19) and Theorem 5.8 (5.31), (5.32),

$$(\mathcal{D})\Sigma|h(I,x)|\chi(X;x) = (Q)\Sigma|h(I,x)| = \sum_{j=1}^{m} (Q_j)\Sigma|h(I,x)| < \sum_{j=1}^{m} V(h;\delta_j;E_j;X_j)$$

$$< \sum_{j=1}^{m}\{V(h;E_j;X_j) + \varepsilon.2^{-j}\} < \sum_{j=1}^{m} V(h;E_j;X_m) + \varepsilon \leq V(h;E;X_m) + \varepsilon \leq \lim_{j\to\infty} V(h;E;X_j) + \varepsilon,$$

giving the opposite inequality to (5.36) and the result for a monotone increasing sequence. For an arbitrary sequence (X_j), as $\bigcap_{j=m}^{\infty} X_j$ is monotone increasing in m,

$$V(h;E;\liminf_{j\to\infty} X_j) = V(h;E;\bigcup_{m=1}^{\infty}\bigcap_{j=m}^{\infty} X_j) = \lim_{m\to\infty} V(h;E;\bigcap_{j=m}^{\infty} X_j) \leq \liminf_{j\to\infty} V(h;E;X_j)$$

as the last set intersection is contained in X_j for every $j \geq m$, so that we can choose suitable j. Thus we prove (5.35). Note that if (X_j) is monotone increasing to X, then (5.35) gives $V(h;E;X) \leq \lim_{j\to\infty} V(h;E;X_j)$, the opposite inequality to (5.36), so that there is equality.

Some results are best seen in one dimension, though they have extensions to R^n. We collect three here, beginning with integration by parts for Stieltjes-type generalized Riemann integrals.

<u>Theorem 5.12</u>: *Let* F,G *be point functions on the closed interval* [a,b] *with* b-a *finite. Let* ΔF *and* dF *stand for* $\Delta(F;[u,v)) = F(v)-F(u)$ *and* dΔF, *respectively. If*

$$(5.37)\qquad \int_{[u,v)} FdG + \int_{[u,v)} GdF = \Delta(FG;[u,v))$$

for all $[u,v) \subseteq [a,b)$, *where both integrals exist, then*

$$(5.38)\qquad V(\Delta F\Delta G;[a,b)) = 0.$$

Conversely, if the first integral exists in (5.37) *for all* $[u,v) \subseteq [a,b)$, *with* (5.38), *then the second integral exists and* (5.37) *is true.*

<u>Proof</u>: We use the identity

$$F(t)\{G(t)-G(w)\} = F(t)G(t)-F(w)G(w)-G(t)\{F(t)-F(w)\} + \{F(t)-F(w)\}\{G(t)-G(w)\}$$

for $t = u$, $w = v$, and for $t = v$, $w = u$, with (5.37). Then

$$V(\Delta F\Delta G;[a,b)) = V(F\Delta G + G\Delta F-\Delta(FG);[a,b)) = V(\int FdG + \int GdF-\Delta(FG);[a,b)) = 0.$$

Conversely, if in (5.37) the first integral exists with (5.38) true, the identity gives

$$V(\int FdG + G\Delta F - \Delta(FG);[a,b)) = 0.$$

As $\Delta(FG) - \int FdG$ is finitely additive, the first definition of variational integral just after Theorem 5.3 shows that $G\Delta F$ is integrable to $\Delta(FG) - \int FdG$ on the partial intervals of $[a,b)$ and (5.37) holds.

To put this into a possibly more recognisable form we can substitute into (5.37),

$$F(v) - F(u) = \int_{[u,v)} fdx, \quad G(v) - G(u) = \int_{[u,v)} gdx,$$

$$(5.39) \qquad \int_{[u,v)} Fgdx + \int_{[u,v)} Gfdx = \Delta(FG;[u,v)),$$

in the sense that (5.39) holds if the integrals F, G and the integrals in (5.39) exist; and if the integrals F,G and the first integral in (5.39) exist, then the second integral there exists with (5.39). These results follow by using Theorem 5.7 (5.25), the integration by substitution, and by noting that (5.38) is automatically true. For by choice of the positive function δ and given $\varepsilon > 0$, we can arrange that

$$|f(x)g(x)|\delta(x) < \varepsilon, \; V(\Delta F\Delta G;\delta;[a,b)) = V(f(x)g(x)(v-u)^2;\delta;[a,b))$$

$$\leq V(\varepsilon(v-u);\delta;[a,b)) = \varepsilon(b-a).$$

As true for all $\varepsilon > 0$, (5.38) is true, and in the calculus form of the integration by parts we do not need (5.38).

The formula (5.37) can be rearranged to a form that can be generalized to n dimensions

$$(5.40)\qquad \int_{[u,v)} \{F(x)-F(u)\}dG = \int_{[u,v)} \{G(v) - G(y)\}dF,$$

where y is the end of the interval I opposite to x. This follows from the identity without using (5.38).

In n dimensions let F be a point function on $J^c = [a,b]$ where $a = (a_1,\ldots,a_n)$, $b = (b_1,\ldots,b_n)$, $a_j < b_j$ $(j = 1,\ldots,n)$, and for $u = (u_1,\ldots,u_n)$, $v = (v_1,\ldots,v_n)$ in J^c with $u_j < v_j$ $(j = 1,\ldots,n)$ let us write

$$\Delta_k F \equiv \Delta_k(F;x,u,v) = F(x_1,\ldots,x_{k-1},v_k,x_{k+1},\ldots,x_n)$$

$$-F(x_1,\ldots,x_{k-1},u_k,x_{k+1},\ldots,x_n),$$

$$\Delta F \equiv \Delta(F;u,v) = \Delta_1\Delta_2 \ldots\Delta_n F.$$

This last is the n-fold difference of F over $[u,v)$.

The repeated integral obtained by integrating over $[a_1,b_1)$, then over $[a_2,b_2),\ldots,$ then over $[a_n,b_n)$, is written using the symbolism $\int\ldots\int_{[a,b)}\cdot\cdot$ This is usually equal to the integral over J, see Chapter 6.

<u>Theorem 5.13</u>: *If the two repeated generalized Riemann integrals in* (5.41) *exist then*

$$(5.41)\qquad \int\ldots\int_{[a,b)} \Delta(F;a,x)dG = \int\ldots\int_{[a,b)} \Delta(G;y,b)dF,$$

where F *and* G *are point functions on* J^c. *The second integral is equal to*

$$(5.42)\qquad \int\ldots\int_{[a,b)} \Delta(G;x,b)dF \qquad \textit{if}$$

$$(5.43)\qquad V(|\Delta(G;y,b) - \Delta(G;x,b)|\Delta(F;x,y);[a,b)) = 0.$$

<u>Proof</u>: Use (5.40) repeatedly.

Theorem 5.12 shows that the converse is easy when $n = 1$, but it is not obvious when $n > 1$, since in (5.43) the difference of two ΔG is a function of b.

Next we look at the Cauchy and Harnack extensions of the generalized Riemann integral. The Cauchy extension was first used on the Riemann integral to integrate some functions unbounded in the neighbourhood of a finite number of points, see Theorems 2.12 to 2.16. In turn, the Lebesgue integral was similarly extended, to integrate some functions that are not summable (Lebesgue integrable) in the neighbourhood of certain points. Another extension, first given by Harnack, extends the integral to an open set if the integral exists, possibly by Cauchy extensions, over each component interval of the open set and if another condition is satisfied. These conditions were first given in one dimension. Denjoy used both extensions repeatedly in a transfinite inductive process in order to integrate all derivatives, and beginning with the Lebesgue integral over certain sets. The theory is tremendously complicated, and only a dedicated student could hope to understand all its details. However, Perron defined a much simpler integral, and in the 1920's it was shown that the Perron and (special) Denjoy integrals are equivalent, they integrate exactly the same functions, to the same values. Now it is even simpler. Our generalized Riemann integral integrates all derivatives (see (3.16)), so that it is not surprising that the Cauchy and Harnack extensions are already included in the generalized Riemann integral.

We first note a uniform continuity property of $h(I,x) - H(I)$ if h integrates to H over all partial sets of the elementary set E. For, taking a single (I,x) out of $\mathcal{D}$ in Theorem 5.3 (5.6), we see that

(5.44) *given* $\varepsilon > 0$, *there is a positive function* δ *on* E^c *such that if* (I,x) is δ-*fine*, $|h(I,x) - H(I)| < \varepsilon$.

In one dimension, taking $x = b$ (see *Ex.* 4.1) and $I = [v,b) \subseteq [u,b)$,

$$h([v,b),b) + H(u,v) - H(u,b) \to 0, \quad h([v,b),b) + H(u,v) \to H(u,b).$$

Thus we are led to the next theorem.

Theorem 5.14: (*Cauchy extension*) *For each* u,v *in* $a \leq u < v < b$ *suppose that* $h(I,x)$ *is integrable over* $[u,v)$ *to* $H(u,v)$. *If there exists*

$$H_1(u,b) \equiv \lim_{v \to b-} [H(u,v) + h([v,b),b)]$$

then $H(a,b)$ *exists and is equal to* $H_1(a,b)$.

Proof: For a strictly increasing sequence (v_j) tending to b- as $j \to \infty$, with $v_0 = a$, then $H(v_{j-1},v_j)$ exists for $j = 1,2,\ldots$ Given $\varepsilon > 0$, there are positive functions δ_j on $[v_{j-1},v_j]$ such that if $\mathcal{D}_j$ is a δ_j-fine division of $[v_{j-1},v_j)$,

$$(5.45) \quad |(\mathcal{D}_j)\,\Sigma\, h(I,x) - H(v_{j-1},v_j)| < \varepsilon.2^{-j} \quad (j = 1,2,\ldots)$$

By definition of $H_1(u,b)$ and the finite additivity of H, if $|v-b| < \delta_0(b)$, for some $\delta_0(b) > 0$,

$$(5.46) \qquad |h([v,b),b) - H_1(v,b)| < \varepsilon.$$

We now take $\delta(b) \leq \delta_0(b)$, $\delta(a) \leq \delta_1(a)$, $\delta(a) \leq v_1-a$, $\delta(x) \leq \min(\delta_j(x), x-v_{j-1}, v_j$ $(v_{j-1} < x < v_j)$, $\delta(v_j) \leq \min(\delta_j(v_j), \delta_{j+1}(v_j), v_j-v_{j-1}, v_{j+1}-v_j)$ $(j = 1,2,\ldots)$

If $\mathcal{D}$ is a δ-fine division of $[a,b)$ we can then split up $\mathcal{D}$ into divisions $\mathcal{D}_j$ of $[v_{j-1},v_j)$ $(j = 1,2,\ldots,J)$, for some integer J, a division $\mathcal{D}_{J+1}$ of $[v_J,v)$ for some v in $v_J < v \leq v_{J+1}$, and $([v,b),b)$. By (5.45), (5.46), and (5.5) in the proof of Theorem 5.1, if $\mathcal{D}_0 = \mathcal{D}\setminus([v,b),b)$,

$$|(\mathcal{D}_0)\Sigma(h-H)+h([v,b),b)-H_1(v,b)| < \sum_{j=1}^{J+1} (\mathcal{D}_j)\Sigma(h-H) + \varepsilon < \sum_{j=1}^{J} \varepsilon.2^{-j} + \varepsilon.2^{-J} + \varepsilon = 2$$

By the finite additivity of H,

$$|(\mathcal{D})\,\Sigma\, h - H(a,v) - H_1(v,b)| < 2\varepsilon, \quad H(a,v) + H_1(v,b) = H_1(a,b),$$

and $H(a,b)$ exists equal to $H_1(a,b)$.

Similarly we can deal with intervals whose left-hand end-points tend to a+.

If we now assume a conventional value $h([v,+\infty), +\infty) = 0$, the proof of Theorem 5.14 suggests a definition of the integral over $[a, +\infty)$. We can use divisions of $[a,+\infty)$ that consist of $([N,+\infty), +\infty)$ for some large N, together with a division of $[a,N)$, and we can let $N \to +\infty$ while δ shrinks, for some positive function δ on $[a,+\infty)$. Taking $h(I,+\infty) = 0$, we have a sum for such a division, and the definition of the integral follows as usual. Theorem 5.14 with b replaced by $+\infty$, shows that the new definition is equivalent to letting $b \to +\infty$ in the integral over $[a,b)$. The importance of the new definition is that the proofs of results such as the various limits under the integral sign go through for $[a,+\infty)$ just as they do for $[a,b)$, we do not have to use a further limit on the result. Similarly Theorem 5.11 (integration by parts) goes through.

Similarly divisions of $(-\infty,b)$ consist of $((-\infty,M),-\infty)$ together with a division of $[M,b)$, with $h(I,-\infty) = 0$, and here $-M$ is large and tends to $+\infty$. Also divisions of $(-\infty, +\infty)$ consist of $((-\infty,M), -\infty)$ and $([N, +\infty), +\infty)$ and divisions of $[M,N)$, with $h(I,x) = 0$ when x is replaced by $-\infty$ and $+\infty$, and $-M$ and N tend independently to $+\infty$. The tests given to Theorem 2.12, 2.13, 2.15-2.19 now have a wider application, to ensure that various generalized Riemann integrals exist.

The Harnack extension is as follows. For $F \subseteq I = [u,v)$, $u \in F$, where F is compact, the open set $G = I \setminus F$ is the union of a sequence (I_k) of disjoint open intervals. If a function f is integrable by some reasonable means over F and the separate I_k, and if

$$(5.47) \qquad \sum_{k=1}^{\infty} M_k$$

is convergent, where

$$M_k = \max_J \left| \int_J f dx \right|$$

for all intervals $J \subseteq I_k$, then the Harnack-extended integral over I, is the sum of the integrals over F and the separate I_k.

To show what this implies, we note that the convergence of (5.47) means that, given $\varepsilon > 0$, there is a K with

$$(5.48) \qquad \sum_{k=K}^{\infty} M_k < \varepsilon ,$$

i.e. if $J_k \subseteq I_k$ for a finite number of distinct $k \geq K$, then

$$(5.49) \qquad \left|\Sigma \int_{J_k} f dx\right| < \varepsilon .$$

Conversely, if (5.49) is true and if we take first the positive integrals and then the negative integrals, we have

$$\Sigma \left|\int_{J_k} f dx\right| \leq 2\varepsilon,$$

for any finite number of distinct $k \geq K$, and so for all $k \geq K$. Hence (5.48) with 2ε for ε, and so (5.47) is convergent. Thus the following theorem shows that the generalized Riemann integral includes the Harnack extension process.

Theorem 5.15: *For* $F \subseteq [a,b]$ *with* $a,b \in F$, *compact, let* $G = (a,b)\setminus F$, *an open set and so the union of disjoint intervals* $[u_j,v_j)$ *with* $[u_j,v_j] \subseteq G$ $(j = 1,2,\ldots)$. *Given* $h(I,x)$, *let*

$$(5.50) \qquad \int_{[a,b)} \chi(F;.)dh, \int_{[u_j,v_j)} dh$$

exist. If, given $\varepsilon > 0$, *there are an integer* J *and a positive function* δ *on* $[a,b]$, *such that for every finite collection* Q *of disjoint intervals* $[u,v) \subseteq [a,b)$, *with* $[u,v) = [u_j,v_j)$ *or* $[u_j,v_j) \cap I$, *for some* $j \geq J$, *some* δ*-fine* (I,x), *and some* $x \in F$, *and no two intervals* $[u,v)$ *lying in the same* $[u_j,v_j)$, *we have*

$$(5.51) \qquad \left|(Q)\Sigma \int_{[u,v)} dh\right| < \varepsilon,$$

then there exists

$$(5.52)\qquad \int_{[a,b)} dh = \int_{[a,b)} \chi(F;.)dh + \sum_{j=1}^{\infty} \int_{[u_j,v_j)} dh.$$

Proof: By Theorem 4.4 (4.4) we can subtract the first integral in (5.50) from both sides of (5.52), and so we need only prove that

$$(5.53)\qquad \int_{[a,b)} \chi(G;.)dh = H_2([a,b)) \text{ where } H_2(E) = \sum_{j=1}^{\infty} \int_{[u_j,v_j)\cap E} dh$$

for each elementary set $E \subseteq [a,b)$. As E is a finite union of intervals each of which has two frontier-points, $[u_j,v_j) \cap E$ is either empty or is $[u_j,v_j)$, for all but a finite number of j. Hence by (5.51) the sequence of partial sums of the infinite series for $H_2(E)$ is fundamental and so convergent, and H_2 exists.

For equality in (5.53) we note that there are $\delta_j(x) > 0$ on $[u_j,v_j]$ such that

$$(5.54)\qquad (\mathcal{D}_j)\, \Sigma\, |h(I,x) - H_2(I)| < \varepsilon.2^{-j},$$

for all δ_j-fine divisions $\mathcal{D}_j$ of $[u_j,v_j)$. Now let $\delta(x) \leq \min(\delta_j(x), v_j-x, x-u_j)$ in $u_j < x < v_j$. Let $\delta(u_j) \leq \delta_j(u_j)$ and $\delta(u_j) \leq \delta_k(v_k)$ if $v_k = u_j$ for some $k (j = 1,2,\ldots)$. As $[u_j,v_j] \subseteq G$, if $x \in F$ then x is not in $[u_j,v_j]$. Thus, for J as in (5.51), we keep $\delta(x) > 0$ so small that $(x-\delta(x), x+\delta(x))$ does not intersect $[u_j,v_j]$ with $j < J$. If $\mathcal{D}$ is a δ-fine division of $[a,b)$ let $(I,x) \in \mathcal{D}$. If $x \in F$ then $\chi(G;x)h(I,x) = 0$ and I can only overlap with $[u_j,v_j)$ if $j \geq J$. Thus the sum of the $H_2(I)$ for such I, will be an infinite series, the limit of a sequence of sums over certain Q satisfying (5.51), and the modulus will not be greater than ε. If, on the other hand, $(I,x) \in \mathcal{D}$, $x \in G$, then for one or possibly two integers j, $x \in [u_j,v_j]$, and (I,x) is δ_j-fine. By the construction of δ it follows that for all $j < J$ the $[u_j,v_j)$ are divided by $(I,x) \in \mathcal{D}$ with $x \in G$, and a finite number of $[u_j,v_j)$ with $j \geq J$ are divided wholly or partially, and using (5.54),

$$|(\mathcal{D})\, \Sigma\chi(G;x)h(I,x) - H_2([a,b))| < \sum_{j=1}^{J-1} \varepsilon.2^{-j} + 2\sum_{j=J}^{\infty} \varepsilon.2^{-j} + \varepsilon < 3\varepsilon,$$

which proves the theorem.

Combining Theorems 3.4, 5.13, 5.14, we see that at each stage of construction of the special Denjoy integral from the Lebesgue integral, the construction is a generalized Riemann integral. Hence the special Denjoy integral is included in the generalized Riemann integral. The latter is wider as it can integrate some interval functions as well.

Ex. 5.1. Let h_1, h_2 be brick-point functions on E^c and α, β be constants. Then

$$V(\alpha h_1 + \beta h_2;E) \leq |\alpha|V(h_1;E) + |\beta|V(h_2;E).$$

Ex. 5.2. If in *Ex.* 5.1 h_1, h_2 are real-valued and $h = h_1 + ih_2$ then

$$V(h_j;E) \leq V(h;E) \ (j = 1,2).$$

Ex. 5.3. If on E^c, h_1 is variationally equivalent to h_2 and h_2 is variationally equivalent to h_3, then h_1 is variationally equivalent to h_3.

Ex. 5.4. If $X \subseteq X_1$, $E \subseteq E_1$, prove that if h is a brick-point function on E^c,

$$V(h;E;X) \leq V(h;E_1;X_1).$$

Ex. 5.5. If E_1, E_2 are partial sets of E with $E_1{}^c$ and $E_2{}^c$ disjoint, and if h is a brick-point function on E^c, then

$$V(h;E;E_1{}^c) + V(h;E;E_2{}^c) = V(h;E;E_1{}^c \cup E_2{}^c).$$

Hint: There are disjoint open sets G_1, G_2 with $G_j \supseteq E_j$ (j = 1,2). Now use *Ex.* 4.2.

Ex. 5.6. Let E_1 be a proper partial set of E and $E_2 = E \setminus E_1$. If $X \cap E_2{}^c$ and $Y \cap E_1{}^c$ are empty, then for a brick-point function h on E^c,

$$V(h;E;X) + V(h;E;Y) = V(h;E;X \cup Y).$$

Ex. 5.7. Let (I_j^o) be a finite or infinite sequence of interiors of disjoint bricks with union J. Then

$$V(h;E;J) = \sum_{j=1}^{\infty} V(h;E;I_j^o).$$

Hint: The left side is not greater than the right, by Theorem 5.6. Now take a finite sum and see *Ex.* 5.6.

Ex. 5.8. In R^1 let $a < b$, $[a,b) \subseteq E$, and $\lim_{\varepsilon\to 0+} V(h;E;[a-\varepsilon, a+\varepsilon]) = 0$. Then

$$V(h;E;[a,b)) = V(h;E;(a,b)).$$

On the other hand, if $\lim_{\varepsilon\to 0+} V(h;E;[b-\varepsilon, b+\varepsilon]) = 0$ then

$$V(h;E;[a,b)) = V(h;E;[a,b]).$$

If both limits are 0 then $V(h;E;X) = V(h;[a,b))$ when $X = (a,b),(a,b],[a,b)$, and $[a,b]$.

Ex. 5.9. If h is a brick-point function on E^c and if $|h|$ is finitely subadditive (in particular, if h is finitely additive) then $|h(I)| \le V(h;I)$, which may be $+\infty$. If $|h|$ is finitely superadditive then h is of bounded variation and $|h(I)| \ge V(h;I)$. If $|h|$ is finitely additive, then h is of bounded variation and $|h(I)| = V(h;I)$. In particular, if $V^*(J) = V(h;J)$ $(J \subseteq I)$, prove that $V(h;I) = V(V^*;I)$.

Ex. 5.10. Let the brick-point function h be of bounded variation in E and, given $\varepsilon > 0$, let the positive function δ on E^c be such that

$$V(h;\delta;E) < V(h;E) + \varepsilon.$$

Then there is a δ-fine division $\mathcal{D}$ of E such that every subdivision $\mathcal{D}_1 \subseteq \mathcal{D}$ satisfies

$$(\mathcal{D}_1) \sum |h| > V(h; \bigcup_{\mathcal{D}_1} I) - 2\varepsilon,$$

Hence for fixed $x \in E^c$ and δ-fine (I,x),

$$\limsup_{\delta(x) \to 0+} |h(I,x)| = \limsup_{\delta(x) \to 0+} V(h;I).$$

Hint: Choose $\mathcal{D}$ so that

$$(\mathcal{D}) \Sigma |h| > V(h;\delta;E) - \varepsilon \geq V(h;E) - \varepsilon.$$

Then by Theorem 5.8 (5.31), (5.32),

$$(\mathcal{D}_1) \Sigma |h| = (\mathcal{D}) \Sigma |h| - (\mathcal{D} \setminus \mathcal{D}_1) \Sigma |h|$$

$$> V(h;E) - \varepsilon - V(h;E \setminus \bigcup_{\mathcal{D}_1} I) - \varepsilon = V(h; \bigcup_{\mathcal{D}_1} I) - 2\varepsilon.$$

Now take an $x \in E^c$. By the construction of *Ex.* 4.1 and since $V(h;E)$ is the infimum of the $V(h;\delta;E)$ we can assume that $\delta(y) < |y-x|$ $(y \in E^c,\ y \neq x)$, so that $\mathcal{D}$ has to contain at least one (I,y) with $y = x$. Let (δ_n) be a sequence of such δ with $\delta_n(x) \to 0+$ and with $\varepsilon = \varepsilon_n \to 0+$. If (I_n,x) is δ_n-fine and in a suitable $\mathcal{D}_n$,

$$|h(I_n,x)| > V(h;I_n) - 2\varepsilon_n,$$

$$\lim_{n\to\infty} |h(I_n,x)| \geq \lim_{n\to\infty} V(h;I_n).$$

The reverse inequality is easy and the proof can be completed using these notes.

Ex. 5.11. (*Cantor's ternary set*) On the real line let $h([u,v),x) = v-u$ and let $[a,b] \subseteq [A,B]$. Prove that $V(h;[A,B);[a,b]) = b-a$ and give a similar result for a finite number of disjoint closed intervals. From $C_0 = [0,1]$ we remove the middle third open interval $(\frac{1}{3}, \frac{2}{3})$ to get C_1. We then remove the middle third open interval of each closed interval in C_1 to get C_2, and we repeat this process to have a sequence (C_j) of sets each consisting of a finite number of disjoint closed intervals. Prove that $V(h;[0,1);C_j) = (2/3)^j$. The intersection C of all the C_j is called *Cantor's ternary set*.

Find $V(h;[0,1);C)$ and $\int_{[0,1)} \chi(C;.)f(.)dh$ where $\chi(C;.)$ is the indicator of C and f is an arbitrary function of points.

(New University of Ulster 1983, M313)

6 The Integrability Of Functions Of Brick-point Functions

To deal with this subject in Lebesgue theory, in which the brick-point functions are of the type $f_j(x)\mu(I)$, a great use is made of the concept of measurability of the point functions $f_j(x)$. Later we will show that an integrable function $f_j(x)\mu(I)$ is such that $f_j(x)$ is μ-measurable, but here we use a different method, which also deals with functions of the more general $h_j(I,x)$. We could use a continuous functional of h_j for the j in some set J, but for simplicity we use a function $r(x_1,\ldots,x_m) : R^m \to R$ and consider the integrability of $r(h_1,\ldots,h_m)$, given that the $h_1,\ldots,h_m$ are all integrable over the same elementary set.

It seems that r has to be continuous in some sense; here we use two senses. First,

$$(6.1) \qquad |r(y_1,\ldots,y_m) - r(x_1,\ldots,x_m)| \le A_1|y_1-x_1| + \ldots + A_m|y_m-x_m|$$

for constants $A_j > 0$, e.g. $|\partial r/\partial x_j| \le A_j$ $(j = 1,\ldots,m)$. More generally, for each $\varepsilon > 0$ and each $x_1,\ldots,x_m$, we put

$$s(x_1,\ldots,x_m;\varepsilon) = \sup |r(y_1,\ldots,y_m) - r(x_1,\ldots,x_m)|(|y_j-x_j| \le \varepsilon,\ j = 1,\ldots,m).$$

Then the continuity condition is that for an arbitrarily large positive integer p, varying $\varepsilon_q > 0$ $(q = 1,\ldots,p)$, and fixed

$$(6.2) \qquad \sum_{q=1}^{p} x_{jq} \quad (j = 1,\ldots,m),$$

$$(6.3) \qquad \sum_{q=1}^{p} s(x_{1q},\ldots,x_{mq};\varepsilon_q) \to 0 \text{ as } \sum_{q=1}^{p} \varepsilon_q \to 0.$$

Clearly (6.1) implies (6.3), but (6.3) is more general than (6.1).

We also assume that

$$(6.4) \qquad r(x_1 + y_1,\dots,x_m+y_m) \le r(x_1,\dots,x_m) + r(y_1,\dots,y_m)$$

for all $(x_1,\dots,x_m)$, $(y_1,\dots,y_m)$ in R^m.

It may be that there are better conditions on r to give what we require.

Theorem 6.1: *Let* h_j-k_j *be of variation zero in the elementary set* E *for* $j = 1,\dots,m$, *where the* h_j, k_j *are brick-point functions defined in* E^c. *Then*

$$(6.5) \qquad r(h_1,\dots,h_m) - r(k_1,\dots,k_m)$$

has variation zero if either r *satisfies* (6.1) *or the* h_j *are integrable in* E *and* r *satisfies* (6.3).

Proof: Given $\varepsilon > 0$, there are positive functions δ_j on E^c such that

$$(6.6) \qquad V(h_j-k_j;\delta_j;E) < \varepsilon \; (j = 1,\dots,m).$$

The δ that is the least of the δ_j at each point, is still positive and can replace the separate δ_j in (6.6). Then (6.1) gives, for each δ-fine division $\mathcal{D}$ and using (6.6),

$$(\mathcal{D}) \Sigma \, |r(h_1,\dots,h_m) - r(k_1,\dots,k_m)| \le (A_1+\dots+A_m)\varepsilon$$

and (6.5) has variation zero.

If h_j is integrable to H_j for each j let the brick-point pairs of $\mathcal{D}$ be (I_q,x_q) $(q = 1,\dots,p)$ and put $x_{jq} = H_j(I_q)$, $y_{jq} = h_j(I_q,x_q)$. Then (6.2) is fixed at $H_j(E)$ and we can take

$$\varepsilon_q = \sum_{j=1}^{m} V(h_j-H_j;\delta;I_q), \quad \sum_{q=1}^{p} \varepsilon_q \le \sum_{j=1}^{m} V(h_j-H_j;\delta;E) \le m\varepsilon,$$

using Theorem 5.8 (5.29) and (6.6). From (6.3), (6.5) has variation zero when k_j is replaced by H_j $(j = 1,\dots,m)$. As h_j-k_j and h_j-H_j have variation zero, so has k_j-H_j, and H_j is also the integral of k_j. Hence (6.5) has variation zero when h_j is replaced by H_j $(j = 1,\ldots,m)$. Putting the two results together, we finish the proof.

Theorem 6.2: *In Theorem* 6.1 *with the* h_j *integrable, let* r *satisfy* (6.4). *Then* $r(h_1,\ldots,h_m)$ *is integrable if, and only if, for a positive* δ *on* E^c *and all* δ*-fine* $\mathcal{D}$ *over* E,

$$(6.7) \qquad (\mathcal{D})\,\Sigma\, r(h_1,\ldots,h_m)$$

is bounded above.

Proof: By (6.4) the brick function $r(H_1(I),\ldots,H_m(I))$ $(I \subseteq E)$ is finitely subadditive.

Theorem 4.7 then gives the results when H_j replaces h_j, and Theorem 6.1 finishes the proof.

Theorem 6.3: *Let* h_j, k_j *be brick-point functions with* $h_j - k_j$ *of variation zero in* $E^c (j = 1,\ldots,m)$. *Then* $\max(h_1,\ldots,h_m) - \max(k_1,\ldots,k_m)$ *has variation zero. If also each* h_j *is integrable to* H_j *then either and so both of* $\max(h_1,\ldots,h_m)$, $\max(H_1,\ldots,H_m)$ *are integrable if and only if there are a bounded above set* S *and a positive function* δ *on* E^c *such that*

$$(6.8) \qquad (\mathcal{D})\,\Sigma\, h_{j(I,x)}(I,x) \in S$$

for all δ*-fine divisions* $\mathcal{D}$ *of* E *and all choices* $j(I,x)$ *from* $(1,2,\ldots,m)$. *Further,* (6.8) *holds if* h *is integrable over* E *and either*

$$(6.9) \qquad h_j(I,x) \le h(I,x) \ (j = 1,\ldots,m) \text{ for all } \delta\text{-fine } (I,x) \text{ in } E^c, \text{ or}$$

$$(6.10) \qquad h_j(I,x) \ge h(I,x) \ (j = 1,\ldots,m) \text{ for all } \delta\text{-fine } (I,x) \text{ in } E^c.$$

Conversely, if $\max(h_1,\ldots,h_m)$ *is proved integrable, it is a suitable* h, *and* $\min(h_1,\ldots,h_m)$ *is also integrable. When* $h_j(I,x) = f_j(x)k(I,x)$ *with* $k(I,x) \ge 0$, $j(I,x)$ *can be independent of* I.

Proof: $x_j = (x_j-y_j)+y_j \leq |x_j-y_j| + y_j \leq |x_1-y_1| +\ldots+|x_m-y_m| + \max(y_1,\ldots,y_m)$

$$\max(x_1,\ldots,x_m) - \max(y_1,\ldots,y_m) \leq |x_1-y_1| +\ldots+ |x_m-y_m|.$$

Interchanging the x's and y's gives (6.1) with $A_j = 1$. For (6.4), take the maximum of

$$x_j + y_j \leq \max(x_1,\ldots,x_m) + \max(y_1,\ldots,y_m).$$

From Theorems 6.1, 6.2 we can now read off the first results. (6.9) implies (6.8) since

$$\max(h_1,\ldots,h_m) \leq h, \quad (\mathcal{D})\Sigma\, h < H(E) + \varepsilon$$

for all δ-fine divisions $\mathcal{D}$ of E and suitable $\delta(x) > 0$, where H is the integral of h.

If (6. 10) holds, so does (6.9) since

$$h_j \geq h,\ h_j-h \leq \sum_{q=1}^{m} (h_q-h),\ h_j \leq \sum_{q=1}^{m} h_q - (m-1)h.$$

If (6.9) holds, possibly with $h = \max(h_1,\ldots,h_m)$, then

$$h_j \leq h,\ h-h_j \leq \sum_{q=1}^{m} (h-h_q),\ h_j \geq \sum_{q=1}^{m} h_q-(m-1)h,\ \min(h_1,\ldots,h_m) = -\max(-h_1,\ldots,-h_m)$$

Thus the max and min are integrable together. Finally, in (6.8) the best choice of j(I,x) is given by

$$h_{j(I,x)}(I,x) = \max(h_1(I,x),\ldots,h_m(I,x)),$$

If $h_j(I,x) = f_j(x)k(I,x)$ with $k(I,x) \geq 0$,

$$\max(h_1(I,x),\ldots,h_m(I,x)) = k(I,x)\max(f_1(x),\ldots,f_m(x))$$

and the choice of j(I,x) can be made independent of I.

Theorem 6.4: *If* t *is fixed in* $0 < t < 1$ *and if in* E^c, $f(x) \geq 0$, $g(x) \geq 0$, $h(I,x) \geq 0$ *with* fh *and* gh *integrable over* E, *then* $f^t g^{1-t} h$ *is integrable over* E.

Proof: Here, $r(x,y) = x^t y^{1-t}$ and we use Hölder's inequality. See Section 12, for example. Then for $p > 1$, $q > 1$, $p^{-1} + q^{-1} = 1$, $u \geq 0$, $v \geq 0$, $s \geq 0$, $w \geq 0$,

$$uv + sw \leq (u^p + s^p)^{1/p}(v^q + w^q)^{1/q}. \quad x_1{}^t x_2{}^{1-t} + z_1{}^t z_2{}^{1-t} \leq (x_1+z_1)^t (x_2+z_2)^{1-t}$$

$$(x_j \geq 0,\ z_j \geq 0,\ j = 1,2),$$

on putting $t = p^{-1}$, $1-t = q^{-1}$, $x_1 = u^p$, $z_1 = s^p$, $x_2 = v^q$, $z_2 = w^q$, and $-r$ satisfies (6.4). Thus on refinement of partitions sums fall and they are bounded below by 0.

The mean value theorem shows that (6.1) is not satisfied. But (6.3) is satisfied, showing the need of something beyond (6.1). We begin with

$$(6.11)\qquad f(x) \equiv x^t + z^t - (x+z)^t \geq 0 \quad (x \geq 0,\ z \geq 0),$$

proved by

$$f(0) = 0,\ f'(x) = t/x^{1-t} - t/(x+z)^{1-t} \geq 0.$$

Thus we prove that

$$(6.12)\quad (x_1+z_1)^t (x_2+z_2)^{1-t} \leq (x_1{}^t + |z_1|^t)(x_2{}^{1-t} + |z_2|^{1-t})$$

$$= x_1{}^t x_2{}^{1-t} + K \quad (x_j \geq 0,\ y_j = x_j + z_j \geq 0,\ j = 1,2),$$

$$K = |z_1|^t x_2{}^{1-t} + x_1{}^t |z_2|^{1-t} + |z_1|^t |z_2|^{1-t}.$$

Next we prove that

$$(6.13)\quad x_1{}^t x_2{}^{1-t} \leq (x_1+z_1)^t (x_2+z_2)^{1-t} + K \quad (x_j \geq 0,\ y_j = x_j + z_j \geq 0,\ j = 1,2),$$

clearly true if $z_j \geq 0$ $(j = 1,2)$. If $z_1 < 0$, (6.11) gives

$$(6.14)\quad x_1{}^t \leq (x_1 - |z_1|)^t + |z_1|^t = (x_1+z_1)^t + |z_1|^t.$$

When $z_1 < 0$, $z_2 \geq 0$, (6.14) need only be multiplied by x_2^{1-t} to give (6.13). Similarly when $z_1 \geq 0$, $z_2 < 0$. If $z_1 < 0$, $z_2 < 0$, (6.14) gives (6.13) since

$$\begin{aligned} x_1^t x_2^{1-t} &\leq \{(x_1+z_1)^t + |z_1|^t\} \{(x_2+z_2)^{1-t} + |z_2|^{1-t}\} \\ &= (x_1+z_1)^t(x_2+z_2)^{1-t} + (x_1+z_1)^t|z_2|^{1-t} + |z_1|^t(x_2+z_2)^{1-t} \\ &\qquad + |z_1|^t|z_2|^{1-t} \\ &\leq (x_1+z_1)^t(x_2+z_2)^{1-t} + K. \end{aligned}$$

Then (6.12), (6.13) give

$$(6.15) \quad |r(x_1,x_2) - r(y_1,y_2)| \leq K = r(|z_1|,x_2) + r(x_1,|z_2|) + r(|z_1|,|z_2|),$$
$$z_j = y_j - x_j \ (j = 1,2).$$

From (6.15), using Hölder's inequality, (6.3) is true since

$$\sum_{q=1}^{p} s(x_{1q},x_{2q};\varepsilon_q) \leq \sum_{q=1}^{p} \{r(\varepsilon_q,x_{2q}) + r(x_{1q},\varepsilon_q) + r(\varepsilon_q,\varepsilon_q)\}$$

$$\leq r\Big(\sum_{q=1}^{p} \varepsilon_q, \sum_{q=1}^{p} x_{2q}\Big) + r\Big(\sum_{q=1}^{p} x_{1q}, \sum_{q=1}^{p} \varepsilon_q\Big) + \sum_{q=1}^{p} \varepsilon_q$$

which tends to 0 with the last sum when (6.2) is fixed.

Previously, when $m = 1$ I have considered the requirements that r be convex and satisfy (6.1) when integrating $r(f(x))h(I,x)$. It turns out that r is then rather restricted in scope unless f has a bounded range of values. For if r has a second derivative everywhere, convexity implies that that derivative is non-negative, and so the first derivative is monotone increasing, and is bounded because of (6.1). Thus even $r(x) = x^2$ is not included, except for a bounded range. To increase the generality we divide R up into a sequence (X_j) of sets such that $A_1 = 2^j$ in X_j, and still obtain what we need. A similar generalization can be used when $r = r(x_1,\ldots,x_m)$.

Theorem 6.5: *Let the convex function* $r: R \to R$, *let* R *be the union of disjoint sets* X_j $(j = 1,2,\ldots)$ *such that for some positive function* δ^+ *on* R,

(6.16) $\quad |r(y)-r(x)| \le 2^j |y-x| (x \in X_j, |y-x| < \delta^+(x),\ j = 1,2,\ldots),$

let $h \ge 0$ *be integrable to* H*, and let* f *be a point function such that* fh *is integrable to* H_1*. Then* $r(f)h$ *is integrable if and only if the set of sums*

(6.17) $\quad (\mathcal{D})\,\Sigma\, r(f)h$

is bounded above, for some positive function δ^* *on* E^c *and all* δ^**-fine divisions* $\mathcal{D}$ *of* E.

Proof: Let W_j be the set of $x \in E^c$ such that $f(x) \in X_j$, and let δ_j be a positive function of E^c such that

(6.18) $\quad (\mathcal{D})\,\Sigma\, |fH - H_1| < \varepsilon.4^{-j}$

for all δ_j-fine divisions $\mathcal{D}$ of E. By the differentiation of integrals (see Section 15) and disregarding those I with $H(I) = 0$, and a set of H-variation zero,

$$|f(x) - H_1/H| < \delta^+(f(x))$$

for all δ^{++}-fine (I,x) and some positive function δ^{++} on E^c. Now take the positive function $\delta(x) \le \min(\delta^*(x), \delta_j(x), \delta^{++}(x))$ when $x \in W_j$, $j = 1,2,\ldots$ If $\mathcal{D}$ is a δ-fine division of E^c let $\mathcal{D}_j$ be the subset of the $(I,x) \in \mathcal{D}$ with $x \in W_j$. Then

$$(\mathcal{D})\Sigma|r(f)H-r(h_1/H)H| = (\mathcal{D})\Sigma|r(f)-r(H_1/H)|H \le \sum_{j=1}^{\infty} 2^j(\mathcal{D}_j)\Sigma|f-H_1/H|H$$

$$= \sum_{j=1}^{\infty} 2^j(\mathcal{D}_j)\Sigma|fH-H_1| \le \sum_{j=1}^{\infty} \varepsilon 2^j\, 4^{-j} = \varepsilon,$$

using (6.16), (6.18). As H can replace h in (6.17) and the integrability argument, see Theorem 5.7 (5.23), then by Theorem 4.7 it is enough to show that $r(H_1/H)H$ is finitely sub-additive. Let P be a partition of a brick I, formed of bricks J. Assuming $H(I) \ne 0$,

$$H(J)/H(I) \ge 0, \quad (P)\,\Sigma H(J)/H(I) = H(I)/H(I) = 1.$$

Thus by the convexity of r,

$$r(H_1(I)/H(I)) = r(P)\,\Sigma H_1(J)/H(I)) = r(P)\,\Sigma\,(H_1(J)/H(J))(H(J)/H(I)))$$

$$\leq (P)\,\Sigma\, r(H_1(J)/H(J))H(J)/H(I),$$

$$r(H_1(I)/H(I))H(I) \leq (P)\ \ \Sigma\, r(H_1(J)/H(J))H(J),$$

proving the finite subadditivity of $r(H_1/H)H$.

Ex. 6.1. Prove Theorem 5.4 (5.14) using $|h| = \max(h,-h)$ with real-valued h.

Ex. 6.2. Let $f(x)h(I,x)$ be integrable over E, let $|f(x)| \leq g(x)$ and let $g(x)|h(I,x)|$ be integrable over E. Prove that $|f(x)h(I,x)|$ is integrable over E.

Ex. 6.3. For $f(x)$, $h(I,x)$ defined on E^c and α a real constant, let h, $|h|$, fh $|fh|$ all be integrable over E. Prove that $\max(fh,\alpha h)$ and $\min(fh,\alpha h)$ are integrable.

Hint: $(|f| + |\alpha|)|h|$ is integrable.

Ex. 6.4. For real x,y let $r(x,y) = (x^2+y^2)^{\frac{1}{2}} = |x+iy|$. Then

$$\big||x+iy| - |r+is|\big| \leq |(x+iy) - (r+is)| \leq |x-r| + |y-s|,$$

$$|(x+r) + i(y+s)| \leq |x+iy| + |r+is|.$$

Hence this r satisfies (6.1) and (6.4).

Ex. 6.5. Let $a_1 < b_1$, $a_2 < b_2$, and let I be the two-dimensional brick with (a_1,a_2) one vertex and (b_1,b_2) the opposite vertex. Let F be a real-valued or complex-valued point function on R^2 and define

$$G_1(F;I) = \int_{[a_1,b_1)} |F(x;b_2) - F(x,a_2)|dx,$$

$$G_2(F;I) = \int_{[a_2,b_2)} |F(b_1,y) - F(a_1,y)|dy,$$

assuming that these generalized Riemann integrals exist for some brick J and all $I \subseteq J$. If G_1, G_2 are themselves generalized Riemann integrable over a brick J, prove that

$$G(F;I) = \{G_1(F;I)^2 + G_2(F;I)^2 + (b_1-a_1)^2(b_2-a_2)^2\}^{\frac{1}{2}}$$

is also generalized Riemann integrable over J.

Hint: This needs the three dimensional analogue of *Ex*. 6.4, *e.g*. by applying it twice. Also G_1 is finitely subadditive in $[a_2,b_2)$, so that integrability of G_1 implies that certain sums are bounded. Similarly for G_2, so that certain sums for G are bounded.

7 The Variation Set

Given an elementary set E and a positive function δ on E^c, let Q be a finite collection of δ-fine brick-point pairs (I,x) with mutually disjoint I lying in E. Let h(I,x) be defined in E^c. Then Q is called a *δ-fine partial division of* E and

$$(Q)\,\Sigma\, h(I,x)$$

a *δ-fine partial sum in* E. The collection of values of all such sums with fixed E, including the empty Q with conventional sum 0, is denoted by $VS(h;\delta;E)$, *the variation set of* h *on* E, *restricted by* δ. If $X \subseteq R^n$ and $\chi(X;x)$ is the indicator of X, the *variation set* $VS(h\chi(X;.);\delta;E)$ *of* h *on* X *and* E, *restricted by* δ, is denoted by $VS(h;\delta;E;X)$. The intersection $VS^c(h;E)$ of closures $VS(h;\delta;E)^c$ for all positive functions δ on E^c, contains 0, and is called *the limiting variation set of* h *on* E, and $VS^c(h;E;X)$ stands for $VS^c(h\chi(X;.);E)$, *the limiting variation set of* h *on* X *and* E. As a division of a partial set E_1 of E, is a partial division of E, a definition from Section 4 gives

(7.1) $S(h;\delta;E_1) \subseteq VS(h;\delta;E)$ $(E_1 \subseteq E)$.

The definitions can hold in value spaces that do not have a modulus or norm with relevant properties, and then the variation set can partially substitute for the variation. In any case it is useful to know how far we can go with variation set theory.

When the value space has a modulus or norm $\|h\|$, then

$$\left\| \sum_{q=1}^{p} h_q \right\| \le \sum_{q=1}^{p} \|h_q\| ,$$

with $VS(h;\delta;E)$ in the closed sphere, centre the origin and radius $V(h;\delta;E)$, $VS^c(h;E)$ lying in the closed sphere, centre the origin and radius $V(h;E)$. However, we can be twice as accurate by taking more care. We begin with a lemma.

Lemma 7.1: *Let real numbers* $x_1,\ldots,x_p$ *have sum* S, *and* V *the sum of the moduli. Then any subset of the numbers has sum* T *lying between* $\frac{1}{2}(S \pm V)$, *the bounds being attained.*

Proof: The sum of the other numbers is $U = S-T$ and $2T-S = T-U$, $|T-U| \le V$, even for $T = 0$ or S. If all positive x_j and no others are in T then $T-U = V$. If all negative x_j and no others are in T then $T-U = -V$.

Theorem 7.2: *For* H *a real-valued finitely additive function of partial sets* E_1 *of* E, *the set of all such* $H(E_1)$ *is bounded if and only if* H *is of bounded variation on* E, *and then*

(7.2) $U(E) \equiv \sup H(E_1) = \frac{1}{2}(H(E)+V(H;E)) \ge 0$, $L(E) \equiv \inf H(E_1) = \frac{1}{2}(H(E)-V(H;E))$

over all partial sets E_1 *of* E. *Hence as* E *varies,* $U(E)$ *and* $L(E)$ *are finitely additive with*

(7.3) $H(E) = U(E) + L(E)$, $V(H;E) = U(E) - L(E)$.

(7.4) *If* H *is not of bounded variation* $\sup H(E_1) = +\infty$, $\inf H(E_1) = -\infty$.

(7.5) *If* $\varepsilon > 0$ *and a proper partial set* E_1 *of* E *satisfy* $H(E_1) > U(E) - \varepsilon$,

so that U(E) *and* V(H;E) *are finite, then* $L(E_1) > -\varepsilon$, $U(E \setminus E_1) < \varepsilon$.

Result (7.3) is the Jordan decomposition of a finitely additive function of bounded variation of bricks, into its positive variation U(E) and its negative variation L(E), just as in one dimension a function of bounded variation is the difference of two monotone increasing functions.

Proof: Let a partition P be a partition P_1 over a partial set E_1, together with a partition P_2 over $E \setminus E_1$ if E_1 is a proper partial set of E. In Lemma 7.1 let the x_j be the H(I) for $I \in P$. Then as H is finitely additive, S = H(E) and, for P fixed,

$$V = (P)\sum |H(I)|, \text{ so that } \tfrac{1}{2}\{H(E) \pm (P)\sum |H(I)|\}$$

are the attained bounds of $H(E_1)$. Now $|H|$ is finitely subadditive, so that if V(H;E) is finite, by refinement of partitions

$$(P)\sum |H(I)|$$

tends upwards to it, giving the values of U(E) and L(E) in (7.2), these being finitely additive in E since H and V(H;E) are, see Theorem 5.8 (5.31).

If $V(H;E) = +\infty$, $(P)\sum |H|$ and so $\pm H(E_1)$ can be arbitrarily large, giving (7.4).

Finally, from $H(E_1) > U(E)-\varepsilon$ and (7.3), (7.2), we have (7.5) from

$$L(E_1) > U(E)-U(E_1)-\varepsilon = U(E\setminus E_1)-\varepsilon \geqq -\varepsilon,\ U(E\setminus E_1)-\varepsilon < L(E_1) \leqq 0.$$

If the value space has a modulus or norm and the variation defined using $\|h\|$,

$$\|H(E_1)-\tfrac{1}{2}S\| \leqq (P)\sum \|\tfrac{1}{2}H\| \leqq V(\tfrac{1}{2}h;\delta;E),\quad \|\tfrac{1}{2}S\| \leqq V(\tfrac{1}{2}h;\delta;E).$$

Returning to real or complex values, by the convention $0 \in VS^c(h;E)$, and 0 is the only possible limit of $VS(h;\delta;E)$ as δ shrinks in E^c. If 0 is the only point contained in $VS^c(h;E)$ and if the closure of $VS(h;\delta;E)$ is compact for some positive function δ on E^c, we say that h is of variation zero on E. To show that this definition agrees with the previous $V(h;E) = 0$ we use the next theorem.

Theorem 7.3: *For a positive function* δ *on* E^c *and a compact set* C *let*

(7.6) $\quad VS(h;\delta;E) \subseteq C.$

Then, given a sphere G *centre the origin of the value space of real or complex numbers, and of radius* $\varepsilon > 0$, *there is a positive function* δ^* *on* E^c *such that*

(7.7) $\quad VS(h;\delta^*;E) \subseteq VS^c(h;E) + G.$

(7.8) $\quad$ *If* (7.6) *is true with* $VS^c(h;E) = \text{sing}(0)$ *then* $V(h;E) = 0$.

Note that if A,B are sets, A+B denotes the set of a+b for all $a \in A$, all $b \in B$.

Proof: In (7.7) the right side is a union of open sets and so an open set G^*. $F \equiv C \setminus G^*$ closed and lies in a compact set, and so is compact. If $y \in F$ then $y \in \setminus VS^c(h;E)$ so that there are a positive function $\delta_y(x)$ on E^c and an open neighbourhood $G(y)$ of y free from points of $VS(h;\delta_y;E)$. The $G(y)$ cover F, so that a finite number cover F, say $G(y_1),\ldots,G(y_p)$. Let $\delta^* = \min(\delta_{y_1},\ldots,\delta_{y_p}) > 0$. Then $F \cap VS(h;\delta^*;E)$ is empty, proving (7.7).

In (7.8), $VS(h;\delta^*;E) \subseteq G$, so that $S(h;\delta^*;E_1) \subseteq G$ for each partial set E_1 of E, and h integrated to 0 on E_1. Hence $V(h;E) = 0$ by Theorem 5.4 (5.12). The converse is easy.

Corollary: It follows from (7.8) and Theorem 5.3 (converse) that h *is integrable to* H *on the partial sets of* E *if and only if* H *is finitely additive*, $VS(h-H;\delta;E)$ *lies in a compact set for some positive function* δ *on* E^c, *and* $VS^c(h-H;E) = \text{sing}(0)$.

<u>Theorem 7.4</u>: *If* E_1 *is a proper partial set of* E *and* h *a brick-point function on* E^c, *then*

(7.9) $VS(h;\delta;E_1) + VS(h;\delta;E\setminus E_1) \subseteq VS(h;\delta;E)$, $VS^c(h;E_1) + VS^c(h;E\setminus E_1) \subseteq VS^c(h;E)$.

(7.10) *If* (7.6) *is true, there is equality in the second result of* (7.9).

<u>Proof</u>: In (7.9), putting $E_2 = E\setminus E_1$, if $y_j \in VS(h;\delta;E_j)$ $(j = 1,2)$ then $y = y_1+y_2$ is a sum for a δ-fine partial division of E, which gives the first result. For the second we use the continuity of addition, so that if X and Y are sets in the value space, $X^c + Y^c \subseteq (X+Y)^c$.

(7.11) $VS^c(h;E_1) + VS^c(h;E_2) \subseteq VS(h;\delta;E_1)^c + VS(h;\delta;E_2)^c \subseteq$

$$\{VS(h;\delta;E_1) + VS(h;\delta;E_2)\}^c \subseteq VS(h;\delta;E)^c,$$

Taking the intersection for all positive functions δ on E^c, finishes the proof of (7.9).

If (7.6) holds, then as $0 \in VS(h;\delta;E_j)$ $(j = 1,2)$ we can replace E by E_j in (7.6).

<u>Lemma 7.5</u>: *If* $X \subseteq C$, *then* $X^c + Y^c = (X + Y)^c$.

<u>Proof</u>: Let $w \in (X + Y)^c$. If there are no $x \in X$, $y \in Y$ such that $x+y = w$, then there are $(x_j) \subseteq X$, $(y_j) \subseteq Y$ such that $x_j+y_j \in X+Y$ and $w_j = x_j+w_j \to w$. As $(x_j) \subseteq C$, the sequence has a limit-point, say x, and a subsequence of (x_j) tends to x. The same subsequence of $(y_j) = (w_j-x_j)$ tends to w-x, and $x \in X^c$, $w-x \in Y^c$, $w \in X^c + Y^c$, proving the lemma.

Returning to the theorem, let δ_R be the positive function of Theorem 4.6 (4.7) and let $\delta \leq \delta_R$ on $E_1 \cup E_2 = E$. By Lemma 7.5 and the proof of Theorem 4.6, the second and third signs $\subseteq$ can be replaced by =, so that if $w \in VS(h;\delta;E)^c$ for all δ then $w = x_1(\delta) + x_2(\delta)$ where $x_j(\delta) \in VS(h;\delta;E_j)^c \subset C$. The collection of $x_1(\delta)$ for all positive δ on E^c, has a limit-point $x_1 \in VS(h;\delta;E_1)^c$ for all positive δ on E^c and a sequence of $x_1(\delta)$ tends to x_1. By continuity of addition the corresponding sequence of $x_2(\delta)$ tends to

$w-x_1$, in $VS(h;\delta^*;E_2)$ for all positive δ^* on E^c. Hence the result.

For a while we fix h and E and vary the X alone, so that for simplicity we can put VS(X) for $VS^c(h;E;X)$. Let (Y_j) be a sequence of sets in the value space R or C. Then the symbols $\sum_{j=1}^{<\infty} Y_j$ and $\sum_{j=1}^{\infty} Y_j$ are defined as follows.

$$\sum_{j=1}^{<\infty} Y_j \equiv \{\sum_{j=1}^{m} y_j : y_j \in Y_j,\ j=1,\ldots,m, \text{ for } m=1,2,\ldots\},\quad \sum_{j=1}^{\infty} Y_j = \{\sum_{j=1}^{\infty} Y_j\}^c.$$

<u>Theorem 7.6</u>: (7.12) *If* $X \subseteq T \subseteq R^n$ *then* $VS(X) \subseteq VS(T)$.

(7.13) *Let* (X_j) *be a sequence of sets in* R^n *with union* X, *let* (C_j) *be a sequence of compact sets in the value space, and let* (δ_j) *be of positive functions on* E^c.

(7.14) *If* $VS(h;\delta_j;E;X_j) \subseteq C_j$ $(j = 1,2,\ldots)$

(7.15) *then* $VS(X) \subseteq \sum_{j=1}^{\infty} VS(X_j)$.

Results (7.12), (7.15) are the respective analogues of *Ex.* 5.4 and Theorem 5.6.

<u>Proof</u>: (7.12) is trivial. Using it, we can assume that the X_j are mutually disjoint. By (7.14) and Theorem 7.3, given $\varepsilon > 0$, there is a positive function δ_j^* on E^c with

(7.16) $VS(h;\delta_j^*;E;X_j) \subseteq VS(X_j) + G_j$

where G_j is the closed sphere centre 0 and radius $\varepsilon.2^{-j}$. Take

$$\delta(x) \leq \delta_j^*(x)\ (x \in X_j),\ \delta(x) \leq 1\ (x \in \setminus X).$$

If Q is a δ-fine partial division of E let Q_j be that part of Q with $x \in X_j$. As Q has only a finite number of brick-point pairs there is a greatest integer m with Q_m not empty. Then, using (7.16),

$$(Q)\ \Sigma\ h\chi(X;x) = \sum_{j=1}^{m} (Q_j)\Sigma\ h \in \sum_{j=1}^{m} VS(h;\delta_j{}^*;E;X_j) \subseteq \sum_{j=1}^{m} \{VS(X_j) + G_j\} =$$

$$= \sum_{j=1}^{m} VS(X_j) + \sum_{j=1}^{m} G_j \subseteq \sum_{J=1}^{m} VS(X_j) + G_0,$$

$$VS(h;\delta;E;X) \subseteq \sum_{j=1}^{<\infty} VS(X_j) + G_0,\ VS(X) \subseteq F + G_0,\ F \equiv \sum_{j=1}^{\infty} VS(X_j).$$

If $y \in \setminus F$, there is a sphere S centre y and radius some $2\varepsilon > 0$, that is free from points of F and then every point of $F + G_0$ is at most within ε of the boundary of S and so a distance at least ε away from y. Hence $VS(X) \subseteq F$, proving (7.15).

Ex. 7.1 We do not always have $X^c + Y^c = (X+Y)^c$. For let X be the set of negative integers and Y the set of points $n + \frac{1}{2} + 1/n$ $(n = 3,4,\ldots)$ Then $X^c = X$, $Y^c = Y$ and $(X+Y)^c$ contains $\frac{1}{2}$, which is not in $X^c + Y^c$.

Ex. 7.2 (P.J. Muldowney) The analogue for variation sets of Theorem 5.8 (5.32) could be as follows. If $VS(h;\delta;E)$ lies in a fixed compact set for each positive δ on E^c, if G is an open neighbourhood of the origin, and if the positive function δ_1 on E^c satisfies $VS(h;\delta_1;E) \subseteq VS^c(h;E) + G$, then $VS(h;\delta_1;E_1) \subseteq VS^c(h;E_1) + G$ for all partial sets E_1 of E. However, the following example shows that the result is false. Let

$$h([u,v);x) = \begin{cases} v-u\ ([u,v) \subseteq [1,2)), \\ u\quad (0<u<1,\ v=1), \\ 0\quad (\text{otherwise}). \end{cases}\quad 0<\delta(1) = \delta<1,\ 0<\delta(x)<|x-1|\ (x \neq 1)$$

then $VS(h;\delta;[0,2)) = [0,2)$, $VS^c(h;[0,2)) = [0,2]$,

$$VS(h;\delta;[0,1)) = (1-\delta,1) \cup \text{sing}(0),\ VS^c(h;[0,1)) = \text{pair } (0,1).$$

Taking $G = (-\varepsilon,\varepsilon)$ with $0 < \varepsilon < \delta$ we contradict the supposed analogue.

Ex. 7.3. We need the compact set in Theorem 7.3 (7.7). For if in R, $h([0,1/n),x) = n$ $(n = 1,2,\ldots)$ with $h = 0$ otherwise, then $V(h;\delta;[0,1)) = +\infty$, $VS(h;\delta;[0,1))$ is unbounded, and yet $VS^c(h;[0,1)) = \text{sing}(0)$.

CHAPTER 3

LIMIT THEOREMS FOR SEQUENCES OF FUNCTIONS

8 Monotone Convergence

Chapters 1 and 2 lay the foundations on which we build more profound properties of the integral. Exx. 4.8, 4.9 show that in this section we cannot consider the most general monotone sequence of brick-point functions. Instead we use a non-negative brick-point function multiplied by a monotone increasing sequence of point functions. We first have a weak form of the monotone convergence theorem which in Lebesgue theory is due to Levy.

Theorem 8.1: *For* E *an elementary set,* $k(I,x) \geq 0$ *in* E^c, *and* $(f_j(x))$ *bounded and monotone increasing in* j, *so that* $f(x) \equiv \lim_{j\to\infty} f_j(x)$ *exists and is finite, for each* $x \in E^c$, *if* $f_j(x)k(I,x)$ *is integrable over* E *to* $H_j(E)$ $(j = 1,2,\ldots)$ *and if the sequence* $(H_j(E))$ *is bounded above with supremum* $H(E)$, *then* $f(x)k(I,x)$ *is integrable over* E *to* $H(E)$, *i.e.*

$$(8.1) \qquad \int_E \lim_{j\to\infty} f_j(x)dk = \lim_{j\to\infty} \int_E f_j(x)dk.$$

Proof: By Theorems 5.1 and 4.4 (4.5) the integral H_j of f_jk exists, monotone increasing in j, for each partial set of E. As $(H_j(E))$ is bounded with supremum $H(E)$, given $\varepsilon > 0$,

$$(8.2) \qquad H(E) - \varepsilon < H_N(E) \leq H(E)$$

for some integer N. By Theorem 4.4 (4.4) we replace f_j by $f_j - f_1 \geq 0$, so that without loss of generality we assume that $f_j \geq 0$. By Theorem 5.3 (5.6), for a positive function δ_j on E^c,

$$(8.3) \quad (\mathcal{D}) \sum |f_j(x)k(I,x) - H_j(I)| \ < \varepsilon.2^{-j}$$

when the division $\mathcal{D}$ of E is δ_j-fine $(j = 1,2,\ldots)$.

Suppose that a function $m(x)$ taking integer values not less than N, is on E^c. Define $\delta(x) = \delta_{m(x)}(x) > 0$ and let $\mathcal{D}$ be a δ-fine division of E using (I,x), with u and v the least and greatest, respectively, of the corresponding values of m(x). Here it is essential that $\mathcal{D}$ has only a finite number of (I,x) so that v is finite. Then $v \geq u \geq N$. As the integral is finitely additive (Theorem 5.1) and by (8.2), the monotonicity, and (8.3),

$$H(E)-\varepsilon<H_u(E) = (\mathcal{D})\Sigma H_u(I) \leq (\mathcal{D})\Sigma H_{m(x)}(I) \leq (\mathcal{D})\Sigma H_v(I) = H_v(E) \leq H(E),$$

$$(8.4) \qquad H(E) - 2\varepsilon < (\mathcal{D})\Sigma f_{m(x)}(x)k(I,x) < H(E) + \varepsilon.$$

This works for every such $m(x) \geq N$. We use two. First let r(x) be the least integer greater than N for which $f_{r(x)}(x) > 0$. If no such r(x) exists, $f_j(x) = 0 (j = 1,2,\ldots), f(x) = 0$, and we define r(x) to be N. Also $s(x) \geq r(x)$ is the least integer, even if r = N, for which

$$(8.5) \qquad f(x) \geq f_{s(x)}(x) \geq f(x) - \varepsilon f_{r(x)}(x).$$

Let $\delta^1(x) = \delta_{r(x)}(x)$, $\delta^2(x) = \delta_{s(x)}(x)$, $\delta(x) = \min(\delta^1(x), \delta^2(x))$. If $\mathcal{D}$ is a δ-fine division of E then $\mathcal{D}$ is δ^1-fine and δ^2-fine, and by (8.4), the monotonicity of $f_j(x)$ in j, and (8.5),

$$H(E)-2\varepsilon < (\mathcal{D})\Sigma f(x)k(I,x) \leq (\mathcal{D})\Sigma f_{s(x)}(x)k(I,x) + \varepsilon(\mathcal{D})\Sigma f_{r(x)}(x)k(I,x)$$

$$< H(E) + \varepsilon + \varepsilon(H(E) + \varepsilon) = (1 + \varepsilon)(H(E) + \varepsilon).$$

As $\varepsilon \to 0+$, the first and last values tend to H(E) and so fk integrates over E to H(E).

This is a rather more general weak convergence theorem than is proved in Lebesgue theory, where k(I,x) = k(I) is a non-negative measure. When $k(I,x) \geq 0$ is of bounded variation we forget r(x), replacing $f(x) - \varepsilon f_{r(x)}(x)$ by $f(x) - \varepsilon$.

Next is the strong monotone convergence theorem, due to Vitali in Lebesgue form,

Theorem 8.2: *In Theorem* 8.1 *we omit the boundedness of* $(f_j(x))$ *in* j *for each* $x \in E^c$, *proving it true* k*-almost everywhere from the boundedness of* $(H_j(E))$. *If* X *is the set of* x *where the limit does not exist, we put* $f(x) = \lim_{j\to\infty} f_j(x)\chi(\sim X;x)$, *and then* $f(x)k(I,x)$ *is integrable over* E *to* H(E) *i.e.*

$$(8.6)\qquad \int_E \lim_{j\to\infty} f_j(x)\chi(\sim X;x)dk = \lim_{j\to\infty}\int_E f_j(x)dk.$$

Proof: Again we replace f_j by $f_j - f_1 \geq 0$, and so can assume $f_j(x) \geq 0$. For X the set where $(f_j(x))$ is unbounded, and $N > 0$ a fixed integer, let t(x) be the least integer such that

$$(8.7)\qquad f_{t(x)}(x) \geq N\ (x \in X),\ t(x) = 1(x \in \sim X).$$

Using Theorem 5.3 (5.6), given $\varepsilon > 0$ let δ_j be a positive function on E^c such that

$$(8.8)\qquad (\mathcal{D})\ \Sigma|f_j(x)k(I,x) - H_j(I)| < \varepsilon.2^{-j}$$

for each δ_j-fine division $\mathcal{D}$ of E, and put $\delta(x) = \delta_{t(x)}(x) > 0$. A δ-fine division $\mathcal{D}$ of E has only a finite number of (I,x), so the corresponding t(x) have a finite maximum v. By (8.7), (8.8) and the monotonicity,

$$(\mathcal{D})\ \Sigma N.k(I,x)\chi(X;x) \leq (\mathcal{D})\ \Sigma f_{t(x)}(x)k(I,x) \leq (\mathcal{D})\ \Sigma H_{t(x)}(I) + \varepsilon$$

$$\leq (\mathcal{D})\ \Sigma H_v(I) + \varepsilon = H_v(E) + \varepsilon \leq H(E) + \varepsilon,\ V(k;E;X) \leq V(k;\delta;E;X) \leq (H(E)+\varepsilon)/N.$$

As true for all integers $N > 0$, $V(k;E;X) = 0$, and by Theorems 5.7 (5.22) and 8.1,

$$\int_E f_j dk = \int_E f_j\chi(\sim X;.)dk \to \int_E fdk.$$

It is usual to omit $\chi(\sim X;.)$.

We can now obtain information about absolutely convergent infinite series.

Theorem 8.3: *Let* $k(I,x) \geq 0$ *and let* $f_j k$ *and* $|f_j|k$ *be integrable in* $E(j = 1,2,\ldots)$. *If*

$$\text{(8.9)} \qquad \sum_{j=1}^{\infty} \int_E |f_j(x)|dk$$

is convergent, then

$$\text{(8.10)} \qquad f(x) \equiv \sum_{j=1}^{\infty} f_j(x)$$

is absolutely convergent k-*almost everywhere and is integrable over* E *with*

$$\text{(8.11)} \qquad \int_E f(x)dk = \sum_{j=1}^{\infty} \int_E f_j(x)dk.$$

Proof: First we use Theorem 8.2 and the results

$$\int_E \sum_{j=1}^{m} |f_j(x)|dk = \sum_{j=1}^{m} \int_E |f_j(x)|dk \leq \sum_{j=1}^{\infty} \int_E |f_j(x)|dk,$$

given convergent, (8.9). Hence $f_0(x)$ is convergent k-almost everywhere, where

$$f_0(x) \equiv \sum_{j=1}^{\infty} |f_j(x)|,$$

and $f_0 k$ is integrable in E, proving the theorem for the $|f_j|$. If the f_j are real-valued, the theorem is true for the $|f_j| + f_j \geq 0$. Subtracting the result for $|f_j|$ gives the result for real-valued f_j. For complex-valued f_j, by Theorems 4.4, 4.5 we deal with the real and imaginary parts of f_j separately.

An absolute integral is now proved to be absolutely continuous.

Theorem 8.4: *If* $k(I,x) \geq 0$ *in* E^c *with* $f(x)k(I,x)$, $|f(x)|k(I,x)$ *and* $k(I,x)$ *all integrable over* E, *then the integral* F *of* fk *is* k-*absolutely continuous in* E^c *and* E.

Proof: For $N > 0$ let $f_N = \min(|f|,N)$. Then f_N is monotone increasing to $|f|$ as $N \to \infty$, and $f_N k$ is integrable by Theorem 6.3 as Nk is integrable. Hence for $N = 1,2,\ldots$,

$$\lim_{N\to\infty} \int_E f_N dk = \int_E |f| dk$$

by Theorem 8.1, and, given $\varepsilon > 0$, there is an integer N such that

$$\int_E (|f|-f_N)dk < \tfrac{1}{2}\varepsilon.$$

If X is any set with $V(k;E;X) < \varepsilon/(2N)$, Theorem 5.4 (5.9), (5.15) with $|f| = |f| - f_N + f_N$ give

$$V(F;E;X) = V(fk;E;X) = V(|f|k;E;X) \leq V((|f|-f_N)k;E;X) + V(f_N k;E;X)$$

$$\leq \int_E (|f| - f_N)dk + N.V(k;E;X) < \varepsilon.$$

Hence the result.

Ex. 8.1. Let the graph of $f_m(x) \geq 0$ in $[-2,2]$ consist of the two sloping sides of an isosceles triangle with base $\sin(m) \leq x \leq \sin(m) + m^{-3}$ and height m, together with that part of the x-axis lying in $[-2,2]$ but not in the base. Find

$$\int_{[-2,2)} f_m dx. \quad \text{Is} \quad \sum_{m=1}^{\infty} f_m(x)$$

convergent k-almost everywhere in $[-2,2]$, when $k([u,v),x) = v-u$?

(New University of Ulster, 1981, M 313)

Ex. 8.2. For sequences that are not monotone we consider the following example. Let $k([u,v),x) \geq 0$ in E^c, and tend to 0 when x is fixed at u or v and $v-u \to 0$. Let $g(x)$ be of bounded variation in E^c. If in R^1, $f_j(x)$ and $k(I,x)$ satisfy the hypotheses of Theorem 8.2, show that $f_j gk$ is integrable and tends to the integrable fgk as $j \to \infty$, where $f_j \to f$ k-almost everywhere. Note that g can change sign.

Hint: g is the difference of two non-negative monotone increasing functions g_1, g_2 and so $g_\ell k \geq 0$ ($\ell = 1,2$). We are given the existence of the integral H_j of $f_j k$. Prove that it is continuous so that $g_\ell H_j$ and so $g_\ell f_j k$ are integrable. Now apply Theorem 8.2 with k replaced by $g_\ell k \geq 0$. Note the use of Theorem 2.3.

Ex. 8.3. By using Theorem 8.1 prove that

$$\lim_{j\to\infty} \int_{[0,1)} \frac{jx-1}{jx+1}\, dx = \int_{[0,1)} \lim_{j\to\infty} \frac{jx-1}{jx+1}\, dx .$$

and hence show that $j^{-1} \log_e j \to 0$ as $j \to \infty$.

Ex. 8.4. If $k(I,x) \geq 0$ is integrable over E and $f(x)$ a real-valued point function in E^c such that fk and $|f|k$ are integrable over E, prove that $g(p,q;x) \equiv \max(\min(f,q),p)$ is integrable when multiplied by k, where $p < q$ are constants, that $h(p,q;x) \equiv (g(p,q;x)-p)/(q-p)$ is monotone decreasing as $p \to q-$ with fixed q, with limit $\chi(X;x)$ the indicator of the set X where $f(x) \geq q$, and that $\chi(X;x)k(I,x)$ is integrable over E.

Hint: $(|f| + |p| + |q|)k$ is integrable. If $f(x) \geq q$, $h(p,q;x) = 1$. If $f(x) \leq p$, $h(p,q;x) = 0$. Thus as $p \to q-$, $h(p,q;x) \to \chi(X;x)$. Next, if $p < r < q$ and $f(x) \geq r$ then $g(p,q;x) = g(r,q;x)$, $h(r,q;x)-h(p,q;x) = (r-p)(g(p,q;x)-q)/\{(q-r)(q-p)\} \leq 0$. If

$$p < f(x) < r,\ g(p,q;x) = f(x) \text{ and } g(r,q;x) = r,$$

$$h(r,q;x) - h(p,q;x) = (p-f(x))/(q-p) < 0.$$

Now use Theorem 8.1 with a suitable sequence of p.

9 Bounded Riemann Sums And The Majorized (Dominated) Convergence Test

When the value space K of the brick-point functions used in the theory, is not the space of real or complex numbers (R or C) it is usual to consider continuous linear functionals on K with values in R, and try to deduce results for K from results for R. There is then a difficulty in applying, say, the monotone and majorized (dominated) convergence tests as it is often difficult to see what precise conditions in K correspond to the required conditions in R. However, it is easy to transfer theory involving compact sets and Riemann sums from R to a K having suitable algebra and topology, which is one reason why the present section appears. Another reason is that it is near to one set of necessary and sufficient conditions for the sequence of integrals to tend to the integral of the limit function, this last integral existing, see Section 11.

Theorem 9.1: *For* E *an elementary set, and* $f_j(x)$ *real-valued and* $k(I,x) \geq 0$, *both on* E^c, *let* $f_j k$ *be integrable to* $H_j(E)$ *over* E, *let* B *be a real number, and let the positive function* δ *on* E^c *be such that for all* δ*-fine divisions* $\mathcal{D}$ *of* E *and all choices of functions* $j(x)$ *taking positive integer values alone,*

$$(9.1) \qquad (\mathcal{D}) \sum f_{j(x)}(x)k(I,x) \geq B.$$

If $\liminf_{j \to \infty} H_j(E)$ *is finite then* $(\liminf_{j \to \infty} f_j)k$ *is integrable over* E *and*

$$(9.2) \qquad \int_E \liminf_{j \to \infty} f_j dk \leq \liminf_{j \to \infty} \int_E f_j dk = \liminf_{j \to \infty} H_j(E).$$

If (9.1) *is replaced by*

$$(9.3) \qquad (\mathcal{D}) \sum f_{j(x)}(x)k(I,x) \leq B$$

for the same $\delta, \mathcal{D}, j(x)$, *with* $\limsup_{j \to \infty} H_j(E)$ *finite, then* $(\limsup_{j \to \infty} f_j)k$ *is integrable over* E *and*

$$(9.4) \qquad \int_E \limsup_{j \to \infty} f_j dk \geq \limsup_{j \to \infty} \int_E f_j dk = \limsup_{j \to \infty} H_j(E).$$

If $f_j \to g$ (finite) k-*almost everywhere, as* $j \to \infty$, *with* (9.1), (9.3) *replaced by*

$$(9.5) \qquad B \leq (\mathcal{D}) \sum f_{j(x)}(x)k(I,x) \leq C$$

for some real numbers $B < C$ *and for all choices of positive integer valued functions* $j(x)$, *then* gk *is integrable over* E *to* $H(E) = \lim_{j\to\infty} H_j(E)$, *which limit exists and is finite, i.e.*

$$(9.6) \qquad \int_E (\lim_{j\to\infty} f_j)dk = \lim_{j\to\infty} \int_E f_j dk.$$

Proof: From Theorem 6.3 and (9.1) or (9.3) we have the integrability of

$$\min(f_N(x),\ldots,f_Q(x))k(I,x) \quad (Q > N \geq 0).$$

By (9.1) the following exists and is finite, and by Theorem 8.1 it, times k, is integrable over E, namely,

$$\inf_{j\geq N} f_j = \lim_{Q\to\infty} \min(f_N,\ldots,f_Q),$$

monotone decreasing in Q. By Theorem 8.2, as

$$\liminf_{j\to\infty} f_j = \lim_{N\to\infty} \inf_{j\geq N} f_j,$$

monotone increasing in N, the lim inf, multiplied by k, is integrable over E provided that $H \equiv \liminf_{j\to\infty} H_j(E)$ is finite and so $(H_j(E))$ bounded below. For

$$\liminf_{j\to\infty} H_j(E) = \lim_{N\to\infty} \lim_{Q\to\infty} \min_{N\leq j\leq Q} \int_E f_j(x)dk \geq \lim_{N\to\infty} \lim_{Q\to\infty} \int_E \min_{N\leq j\leq Q} f_j(x)dk$$

$$= \lim_{N\to\infty} \int_E \inf_{j\geq N} f_j(x)dk, = \int_E \liminf_{j\to\infty} f_j(x)dk$$

since the previous integral is bounded above by H. Hence (9.2). Then (9.4) follows from (9.3) and $(H_j(E))$ being bounded above, by putting $-f_j$ for f_j in (9.2). For (9.6) from (9.5) we use the other two results, noting that from (9.5), $(H_j(E))$ is bounded, so that

$$\int_E g\,dk = \int_E (\liminf_{j\to\infty} f_j)dk \le \liminf_{j\to\infty} \int_E f_j dk \le \limsup_{j\to\infty} \int_E f_j dk$$

$$\le \int_E (\limsup_{j\to\infty} f_j)dk = \int_E g\,dk, \text{ giving (9.6).}$$

Note that in (9.5) the set of values of sums lies in a compact set, and this is easily translatable to more general value spaces.

We can now prove the majorized (dominated) convergence test.

Theorem 9.2: *Let* $f_j k$, pk, qk *all be integrable in an elementary set* E *with* $k(I,x) \ge 0$, *and* f_j, p, q *all being point functions, in* E^c. *If* $p \le f_j$ *for all* j *with finite*

$$\liminf_{j\to\infty} \int_E f_j(x)dk, \textit{ then } (\liminf_{j\to\infty} f_j)k$$

is integrable over E *and* (9.2) *is true. On the other hand, if* $f_j \le q$ *with finite*

$$\limsup_{j\to\infty} \int_E f_j dk, \textit{ then } (\limsup_{j\to\infty} f_j)k$$

is integrable over E *and* (9.4) *is true. If* $p \le f_j \le q$ *for all* j *and* $f_j \to g$, *finite, k-almost everywhere, as* $j \to \infty$, *then* gk *is integrable and* (9.6) *is true. Note that in Lebesgue integration theory, the first part is called Fatou's lemma.*

Proof: Given $\varepsilon > 0$, there is a positive function δ on E^c such that for all δ-fine divisions $\mathcal{D}$ of E,

$$|(\mathcal{D}) \sum p(x)k(I,x) - \int_E p\,dk| < \varepsilon ,$$

and so in the first part,

$$\int_E p\,dk - \varepsilon < (\mathcal{D}) \sum p(x)k(I,x) \le (\mathcal{D}) \sum f_{j(x)}(x)k(I,x),$$

proving (9.1). Thus Theorem 9.1, first part, gives the result. The rest of the proof is as for Theorem 9.1.

An extension of Theorem 9.2 is as follows.

Theorem 9.3: *Let* $p_j(x) \to p(x)$ *and* $q_j(x) \to q(x)$ *as* $j \to \infty$, k*-almost everywhere in* E^c, *where* $k(I,x) \geq 0$, *such that* p_jk, pk, q_jk, qk *are all integrable over* E *and as* $j \to \infty$,

$$\int_E p_j dk \to \int_E pdk, \quad \int_E q_j dk \to \int_E qdk.$$

Then we can replace p,q *by* p_j,q_j *in Theorem* 9.2.

Proof: In the first part, $p_j \leq f_j$ means that $f_j - p_j \geq 0$. Proving the first part for $f_j - p_j$ we then have it for f_j by adding p_j, since

$$\int_E \liminf_{j \to \infty} (f_j - p_j)dk = \int_E \liminf_{j \to \infty} f_j dk - \int_E pdk,$$

$$\liminf_{j \to \infty} \int_E (f_j - p_j)dk = \liminf_{j \to \infty} \int_E f_j dk - \int_E pdk.$$

The tests involved in Theorems 9.2, 9.3 basically belong to absolute integration, as is shown by Theorem 9.3, in that $f_m - p_m$ or $f_m - f_j$ is an absolute integral. There is a similar restriction of Theorem 9.1, last part.

Theorem 9.4: *If* $f_j(x)$ *and* $k(I,x)$ *satisfy the conditions of Theorem* 9.1, *last part, for* E^c, *then* $(f_j - f_1)k$ *is absolutely integrable on* E.

Proof: Let $\mathcal{D}$ be a δ-fine division of E, where δ is as for (9.5). First take $j(x) = 1$ (all x) and then take $j(x) = j$ when $f_j(x) - f_1(x) \geq 0$, say in X, and $j(x) = 1$ in $\setminus X$. The modulus of the difference of the two sums satisfying (9.5) is at most $C-B$, i.e.

$$(\mathcal{D})\Sigma|f_j(x) - f_1(x)|\chi(X;x)k(I,x) \leq C-B.$$

Similarly

$$(\mathcal{D})\Sigma|f_j(x) - f_1(x)|\chi(\sim X;x)k(I,x) \leq C-B$$

$$(\mathcal{D})\Sigma|f_j(x) - f_1(x)|k(I,x) \leq 2C - 2B,$$

and $(f_j-f_1)k$ is of bounded variation on E. Hence by Theorem 5.4 (5.13), (5.14), $(f_j-f_1)k$ and its modulus are integrable on E.

Clearly we can replace f_j-f_1 by $f_{j(x)}-f_1$ or $f_j-f_{j(x)}$.

Ex. 9.2 shows that sometimes the conclusion of Theorem 9.1 is satisfied without (9.5) being true. Thus (9.5) is not necessary for (9.6).

Ex. 9.1. If $f_j(x) = j/\{1+j^2(jx-1)^2\}$ prove that for all x, $\lim_{j\to\infty} f_j(x) = 0$ and that by the calculus, $\lim_{j\to\infty}\int_{[0,1)} f_j(x)dx = 0$.

If in this example [m] denotes the integer part of m, prove that for $x > 0$,

$$1-x < x[x^{-1}] \leq 1,\ 0 \leq (x[x^{-1}] - 1)^2 < x^2.$$

Hence taking $j = [x^{-1}]$, show that

$$\sup_j f_j(x) > \tfrac{1}{2}[x^{-1}] \geq \tfrac{1}{2}(x^{-1} - 1) \quad (0 < x \leq 1)$$

and that if $g(x) \geq f_j(x)$ for all j then g cannot be integrable in [0,1). (New University of Ulster, 1980, M 313)

Ex. 9.2. Let the finite-valued point function g on E^c be integrable when multiplied by $k(I,x) \geq 0$ over E, but not absolutely integrable. Then as $j \to \infty$, $f_j \equiv g/j$ tends to $f = 0$ at all points of E^c while

$$\int_E f_j dk = \frac{1}{j}\int_E g dk \to 0,$$

and yet $f_j-f_1 = g(1-j)/j$ is not absolutely integrable for $j > 1$. By Theorem 9.4 this example does not satisfy (9.5), so that (9.5) is not necessary for convergence under the integral sign.

Ex. 9.3. Let $n = 1$, $E = [0,1)$, $f_j(x) = x^{-3/2}$ for $(j+1)^{-1} < x \le j^{-1}$ and $f_j = 0$ otherwise. Then $f = 0$ and the integral tends to 0. If $g \ge f_j$ for all j, g is not integrable.

Ex. 9.4. Let $g(x)$ be of bounded variation on E^c and let $f_j(x)$ and $k(I,x)$ on E^c satisfy Theorem 9.1 (third part). Prove that $f_j(x)g(x)$ and $k(I,x)$ satisfy that theorem.

<u>Hint</u>: See Ex. 8.2.

Ex. 9.5. For G a bounded open set on the complex plane with boundary a simple closed contour S traced out by a complex-valued function $z(x)$ when the real variable x goes from 0 to 1, let $(f_j(z))$ be a sequence of functions of z with limit $f(z)$ as $j \to \infty$. Then we can define the integral of f_j round S to be

$$\int_S f_j(z)dz = \int_{[0,1)} f_j(z(x))dz(x).$$

Let $f_j(z(x))z'(x)$ and $k([u,v),x) = v-u$ satisfy Theorem 9.1 (third part). Let $T \subseteq G$ be a closed contour, so that $T \cap S$ is empty, and let

$$g_j(t) = \frac{1}{2\pi i}\int_S \frac{f_j(z)}{z-t}\,dz \quad g(t) = \frac{1}{2\pi i}\int_S \frac{f(z)}{z-t}\,dz.$$

Noting that $(z-t)^{-1}$ is of bounded variation for the $z \in S$ when t is fixed on T, prove that g_j and g exist and are regular (complex variable sense) on T and $g_j \to g$ uniformly for $t \in T$. Thus the full power of limits under the integral sign is only needed on the boundary S, the situation is simpler on T.

<u>Hint</u>: Use Ex. 9.4 and Vitali's convergence theorem from complex variable theory, for a contour between S and T. Vitali's theorem is as follows.

Let $(f_j(z))$ be a sequence of functions, each regular in a region D. For a fixed constant M, every integer j, and every $z \in D$, let $|f_j(z)| \le M$. Further, let $f_j(z) \to f(z)$ as $j \to \infty$, at a set of points z having a limit-point inside D. Then $f_j(z) \to f(z)$ uniformly in any region bounded by a contour interior to D.

10 Controlled Convergence

Another way of dealing in one dimension with limits under the integral sign uses extensions of definitions given in Section 5 above Theorem 5.10. When h depends on a parameter j, we say that h_j is k-AC* in X, uniformly in j, if $V(h_j : E;X_0) \to 0$ uniformly in j when $V(k;E;X_0) \to 0$ and $X_0 \subseteq X$. Then the definition of k-ACG* uniformly in j, follows easily. When k = m and $m([u,v);x) = v-u$, we drop k- from k-AC*, etc.

<u>Lemma 10.1</u>: (101) *Let* $(f_j(x))$ *be a sequence of real or complex valued point functions defined on* [a,b] *that converges almost everywhere in* [a,b] *to a finite* f(x) *as* $j \to \infty$, *where each* $f_j m$ *is integrable to* $F_j(I)$ *on all intervals* $I \subseteq [a,b)$.

(10.2) *Let* F_j *be* AC* *in* [a,b) *uniformly in* j, *so that the* F_j *are absolute integrals,*

(10.3) *Then* $V(F_j - F_p; [a,b)) \to 0$ *as* $j,p \to \infty$.

(10.4) *Further, let* $F_j \to F$ *as* $j \to \infty$.

(10.5) *Then* $V(F - F_p ; [a,b)) \to 0$ *as* $p \to \infty$.

<u>Proof</u>: By Theorem 5.7 we can assume that $f_j \to f$ everywhere in [a,b]. For if $\sim X$ is the set where f_j does not tend to a finite limit, we replace f_j by $f_j\, \chi(X:.)$. Given $\varepsilon > 0$, let Y_p be the set of x where

(10.6) $|f_j(x) - f(x)| < \varepsilon$ (all $j \geq p$).

Then (Y_p) is monotone increasing with limit [a,b], and by Theorem 5.11,

$$V(m;[a,b);Y_p) \to V(m;[a,b))(p \to \infty);\ V(m;[a,b);\sim Y_p) \to 0\ (p \to \infty)$$

by the measurability results of Section 16. Using (10.2),

(10.7) $V(F_j;[a,b);\sim Y_p) < \varepsilon\ (p \geq p_0(\varepsilon))$,

uniformly in j. This uniformity needs (10.2), though for fixed j, Theorem

5.11 gives the result directly. Further,

$$V(F_j-F_p;[a,b)) \leq V(\int (f_j-f_p)dm; [a,b);Y_p) + V(F_j-F_p;[a,b);\sim Y_p)$$

$$\leq V((f_j-f_p)m; [a,b);Y_p) + 2\varepsilon \leq 2\varepsilon(b-a+1) \quad (p \geq p_0(\varepsilon)),$$

giving (10.3). We have used (10.6), (10.7) and Theorems 5.6, 5.4 (5.9). Using (10.4) let $j \to \infty$. As F_j is finitely additive and so $|F_j-F_p|$ is finitely subadditive, then for each division $\mathcal{D}$ of $[a,b)$,

$$(\mathcal{D})\Sigma|F-F_p| = \lim_{j\to\infty} (\mathcal{D})\Sigma|F_j-F_p| \leq \liminf_{j\to\infty} V(F_j-F_p;[a,b)) \leq 2\varepsilon(b-a+1)(p\geq p_0(\varepsilon)),$$

$$V(F-F_p; [a,b)) \leq 2\varepsilon(b-a+1) \quad (p \geq p_0(\varepsilon))$$

and (10.5) follows.

Lemma 10.2: *If we assume* (10.1) *and replace* (10.2), (10.4) *by* (10.8), (10.9),

(10.8) *the* $F_j([a,x))$ *converge uniformly on* $[a,b]$ *to* $F([a,x))$ $(a < x \leq b)$;

(10.9) *the* F_j *are* $AC^*(X)$ *uniformly in* j *where* X *is a closed subset of* $[a,b]$,

(10.10) *then* $V(F_j-F_p; [a,b);X) \to 0$, $V(F-F_p;[a,b);X) \to 0$, *as* $j,p \to \infty$.

Proof: Given the finitely additive function F_j of partial sets of $[a,b]$, we define the point function G_j on $[a,b]$ by $G_j(x) = F_j([a,x))$ $(x \in X)$, and on the intervals (u,v) of $(a,b)\setminus X$ with $u,v \in X$ we take G_j linear from $G_j(u)$ to $G_j(v)$. By (10.8), (10.9) the G_j are absolutely continuous uniformly in j, while for $u \leq r < s \leq v$,

$$G_j(s) - G_j(r) = \int_{[r,s)} g_j dm, \quad g_j(x) = F_j([u,v))/(v-u) \quad (u < x < v),$$

and by (10.8) this g_j tends to $F([u,v))/(v-u)$ everywhere in (u,v). If

$r < s$, both in X,

$$G_j(s) - G_j(r) = F_j([r,s)) = \int_{[r,s)} f_j dm,$$

so that if also $g_j = f_j$ in X, g_j tends to a finite limit almost everywhere. Hence by Theorem 5.4 (5.9) and, Lemma 10.1 (10.3),

$$V(F_j-F_p;[a,b);X) = V(f_jm-f_pm;\ [a,b);X) = V(g_jm-g_pm;\ [a,b);X)$$

$$= V(\Delta G_j-\Delta G_p;\ [a,b);X) \leq V(\Delta G_j-\Delta G_p;[a,b)) \to 0(j,p \to \infty).$$

Similarly the second result follows from Lemma 10.1 (10.5).

<u>Theorem 10.3</u>: *If* (10.1), (10.8) *hold, and*

(10.11) *the functions* F_j *are* ACG* *uniformly in* j,

(10.12) *then* fm *in integrable on* $[a,b)$ *and*

$$\int_{[a,b)} f_j dm \to \int_{[a,b)} fdm\ (j \to \infty).$$

<u>Proof</u>: Let (X_q) be the sequence of closed sets used in the definition of ACG*, let

$$Z_p = \bigcup_{j=1}^{p} X_j,\ W_p = Y_p \cap Z_p;$$

where Y_p is as in Lemma 10.1 (10.6). By Lemma 10.2 let $r(q)$ be the integer such that

(10.13) $V(F_r-F;\ [a,b);X_q) < \varepsilon.2^{-q}$ (all $r \geq r(q)$),

given $\varepsilon > 0$, and take $s = \max(r(1),\ldots,r(p))$. Let δ_q be a positive function on $[a,b]$ such that

(10.14) $V(f_s-F;\delta_q;[a,b);X_q) < V(F_s-F;[a,b);X_q) + \varepsilon.2^{-q} < \varepsilon.2^{1-q}$

and let the positive function δ on [a,b] satisfy $\delta \leq \delta_q$ (q = 1,...,p) and be such that for all δ-fine divisions $\mathcal{D}$ of [a,b).

$$(10.15)\quad (\mathcal{D})\Sigma|f_s(x)(v-u) - F_s([u,v))| < \varepsilon .$$

From (10.6), (10.15), then (10.13), (10.14),

$$(\mathcal{D})\Sigma|f(x)(v-u) - F([u,v))|\chi(W_p;x) \leq (\mathcal{D})\Sigma|f(x)-f_s(x)|(v-u)\chi(Y_p;x) +$$

$$+ (\mathcal{D})\Sigma|f_s(x)(v-u) - F_s([u,v))| + (\mathcal{D})\Sigma|F_s([u,v)) - F([u,v))|\chi(Z_p;x)$$

$$< \varepsilon\,(b-a) + \varepsilon + \sum_{q=1}^{p} V(F_s-F;\delta_q;\ [a,b);X_q) < \varepsilon\,(b-a+3),$$

$$V(fm - F;\ [a,b);W_p) \leq \varepsilon(b-a+3).$$

As $p \to \infty$, W_p is monotone increasing and tends to [a,b], so that by Theorem 5.11,

$$V(fm - F;\ [a,b)) \leq \varepsilon(b-a+3).$$

This is true for all $\varepsilon > 0$, so that fm-F has variation zero on [a,b). Now f is the limit of f_j and F is the limit of F_j, which are finitely additive. Thus F is finitely additive and so is the variational integral of fm, which is equivalent to the (generalized Riemann) integral by Theorem 5.3, giving the final result of Theorem 10.3.

11 Necessary And Sufficient Conditions

In this section we look for necessary and sufficient conditions for a sequence of integrable functions to tend to an integrable limit, and for the integrals of members of the sequence to tend to the integral of the limit function. In more detail, (11.1). *Let* E *be an elementary set, let the real-valued* k(I,x) *of bounded variation be defined in* E^c, *and let* $f_j : E^c \to R$ *with* f_jk *integrable in* E (j = 1,2,...) *such that* $f_j(x) \to f(x)$ *for all* $x \in E^c$, *as* $j \to \infty$.

This basic condition is not too restrictive since if the convergence of f_j to f is only known k-almost everywhere, it is then everywhere except in a set X with $V(k;E;X) = 0$, and we can replace f_j and f by $f_j\chi$ and $f\chi$ where χ is the indicator of the complement of X. The replacement does not alter the integrability of f_jk, fk, nor the values of the corresponding integrals if they exist.

We then look at two properties,

(11.2) *the integrability of* fk *over* E,

(11.3) $$\lim_{j\to\infty}\int_E f_j dk = \int_E f dk = \int_E \lim_{j\to\infty} f_j dk.$$

Theorem 11.1: *Given* (11.1), *a necessary and sufficient condition for* (11.2) *is that*

(11.4) $$(\mathcal{D})\,\Sigma\, f_{m(x)}(x)k(I,x) \in S,$$

for some compact set S *of arbitrarily small diameter, some positive functions* M, δ, *on* E^c, *all positive integer valued functions* $m \geq M$ *on* E^c, *and all* δ*-fine divisions* $\mathcal{D}$ *of* E. *If the diameter condition on* S *is omitted, the weakened condition is not always sufficient.* Note how close (11.4) is to (9.5), which has no diameter condition.

Proof: As $j \to \infty$, f_j tends to f everywhere in E^c. Thus, given $\varepsilon > 0$, there is a positive function M such that

(11.5) $$|f_j(x) - f(x)| < \varepsilon \quad (\text{all } j \geq M(x), \text{ all } x \in E^c).$$

If fk is integrable to F on E, there is a positive function δ on E^c such that

(11.6) $$F-\varepsilon < (\mathcal{D})\,\Sigma\, f(x)k(I,x) < F + \varepsilon$$

for all δ-fine divisions $\mathcal{D}$ of E. Hence using (11.5), (11.6),

$$F-\varepsilon(1+V) < (\mathcal{D})\Sigma f_{m(x)}(x)k(I,x) < F+\varepsilon(1+V) \quad (m(x) \geq M(x),\ V = V(k;\delta;E))$$

and we have (11.4) with S of arbitrarily small diameter,

$$(11.7) \qquad S = [F-\varepsilon(1+V),\ F + \varepsilon(1+V)].$$

Conversely, letting m(x) tend to infinity at each point x, (11.4) gives

$$(11.8) \qquad (\mathcal{D})\ \Sigma\, f(x)k(I,x) \in S$$

as S is compact. If S has arbitrarily small diameter by choice of M, δ, as in (11.7), then (11.8) shows that fk is integrable in E. But if we do not assume the diameter condition on S, then (11.8) is not always sufficient, see Ex. 11.1 and Theorem 16.10.

Theorem 11.2: *Given* (11.1), (11.2), *a necessary and sufficient condition for* (11.3) *is that, given* $\varepsilon > 0$, *there are a positive integer* J *and a positive function* δ_j *on* E^c *and depending on* j, *with*

$$(11.9) \qquad F - \varepsilon < (\mathcal{D})\ \Sigma f_j(x)k(I,x) < F + \varepsilon,\ F \equiv \int_E fdk,$$

for all δ_j*-fine divisions* $\mathcal{D}$ *of* E *and all* $j \geq J$.

Proof: We are given that the integrals F_j and F of f_jk and fk, respectively, over E exist. It is thus necessary and sufficient that, given $\varepsilon > 0$, there are a positive integer J and a positive function δ_j on E^c and depending on j, with

$$F - \tfrac{1}{2}\varepsilon < F_j < F + \tfrac{1}{2}\varepsilon \quad (\text{all } j \geq J),$$

and so (11.9), leading to a compact set $[F-\varepsilon, F+\varepsilon]$ of arbitrarily small diameter.

The necessary and sufficient conditions given in Theorems 11.1, 11.2 are generalized to more general convergence in Section 13, and then specialized to particular cases in Sections 15, 18.

Ex. 11.1. If $f_j = 1$ in [0,1) on the real line, $f_j = (-1)^q q(q+1)$ in

$2-1/q \le x < 2-1/(q+1)$ $(q = 1,2,\ldots,j)$, and $f_j = 0$ in $2-1/(j+1) \le x \le 2$, then f_j is integrable over $[0,2)$. The limit f as $j \to \infty$ satisfies $f = 1$ in $[0,1)$, $f = (-1)^q q(q+1)$ in $2-1/q \le x < 2-1/(q+1)$ $(q = 1,2,\ldots)$, $f(2) = 0$. In the respective intervals take a function M satisfying $M \ge 1$, $M \ge q$, with $M(2) \ge 1$. If $m \ge M$ then $f_m = f$. For suitable $\delta > 0$ the sums oscillate between 0 and 1, while

$$\int_{[0,2-1/j)} f dx = 1 + \sum_{q=1}^{j} (-1)^q$$

and does not tend to a limit as $j \to \infty$, and f is not integrable over $[0,2)$ since the integral has the Cauchy limit property, see Theorem 5.14.

Note that this example does not satisfy (9.5) in Theorem 9.1 since by suitable choice of $j(x)$ the negative terms can be replaced by 0, and the positive terms give unbounded sums.

12 Mean Convergence And L^p Spaces

Let p be a positive real number. Then for integration over the elementary set E in R^n,

$$(12.1) \qquad N(f,g)_p \equiv \int_E |f(x) - g(x)|^p \, dh$$

is a metric when $p \le 1$. When $p > 1$, $N(f,g)_p^{1/p}$ is a metric. These metrics can be used for 'distances' between the functions f and g in suitable spaces of functions, or function spaces, which may explain the usefulness of Theorem 12.1.

<u>Theorem 12.1</u>: *Let* $h(I,x) \ge 0$, *let* $(f_j(x))$ *be a sequence of point functions, both on* E^c, *and let the* $N(f_j,f_q)_p$ *of* (12.1) *exist for every pair* j,q *of positive integers. Suppose that, given* $\varepsilon > 0$, *a* $p(\varepsilon)$ *satisfies*

$$(12.2) \qquad N(f_j,f_q)_p < \varepsilon \text{ (all } j,q \ge p(\varepsilon)).$$

Then there are a point function f *and a subsequence* $j(r)$ *of the positive integers such that*

(12.3) $$f(x) = \lim_{r\to\infty} f_{j(r)}(x),\ \textit{finite, h-almost everywhere in } E^c,$$

and for each positive integer $j \geq p(\varepsilon)$, *the integral below exists and*

(12.4) $$N(f_j,f)_p \leq \varepsilon.$$

(12.5) *If* g *satisfies the condition for* f *in* (12.4) *then* $f = g$ *h-almost everywhere, a uniqueness result.*

The older notation is that a sequence (f_j) satisfying (12.2) is said to *converge in mean with index* p, and one satisfying (12.4) is said to *converge in mean to* f *with index* p. The more modern notation notes that $N(f,g)_p$ or $N(f,g)_p^{1/p}$ is a metric, and calls the first sequence *fundamental* or *Cauchy*, and the second sequence, *convergent to* f, under the given metric. The theorem says that a fundamental sequence is convergent, so that the corresponding function space is *complete*.

Proof of theorem: When $p \neq 1$ we cannot use Theorem 8.3. So we choose a strictly increasing sequence $(j(r))$ with $N(f_j,f_q)_p < 2^{-2rp}$ $(j,q \geq j(r))$. Let X_r be the set where

(12.6) $$|f_{j(r)}(x) - f_{j(r+1)}(x)| \geq 2^{-r}.$$

By Theorem 5.4 (5.13),

$$2^{-rp}V(h;E;X_r) \leq V(|f_{j(r)}(x) - f_{j(r+1)}(x)|^p h(I,x);E)$$

$$= N(f_{j(r)},f_{j(r+1)})_p < 2^{-2rp},\ V(h;E;X_r) < 2^{-rp}.$$

If $x \notin Y_M \equiv \bigcup_{j\geq M} X_j$ then

$$\sum_{r=M}^{\infty} |f_{j(r)}(x) - f_{j(r+1)}(x)| < \sum_{r=M}^{\infty} 2^{-r} = 2^{-M+1}.$$

and there exists

$$f_{j(1)}(x) + \sum_{r=1}^{\infty} \{f_{j(r+1)}(x) - f_{j(r)}(x)\} = \lim_{r\to\infty} f_{j(r)}(x), = f(x),$$

say. Thus $f(x)$ exists in $\setminus X$ where

$$X = \bigcap_{M=1}^{\infty} Y_M,\ V(h;E;X) \le V(h;E;Y_M) \le 2^{-Mp}/(1-2^{-p}),\ V(h;E;X) = 0.$$

Thus f exists h-almost everywhere and Fatou's lemma (Theorem 9.1 (9.2)) gives

$$\int_E |f_j(x) - f(x)|^p\, dh \le \liminf_{r\to\infty} \int_E |f_j(x) - f_{j(r)}(x)|^p\, dh \le \varepsilon(j \ge p(\varepsilon)),$$

and (12.3), (12.4). If also, for every $\varepsilon > 0$,

$$\int_E |f_j(x) - g(x)|^p\, dh \le \varepsilon\ (j \ge q(\varepsilon)),$$

then Fatou gives

$$\int_E |f(x) - g(x)|^p\, dh \le \liminf_{r\to\infty} \int_E |f_{j(r)}(x) - g(x)|^p\, dh \le \varepsilon,$$

$$\int_E |f(x) - g(x)|^p\, dh = 0,$$

and Theorems 5.4 (5.13) and 5.7, converse of the first part, give

$$V(|f(x) - g(x)|^p\, h(I,x);E) = 0,\ f(x) = g(x)$$

h-almost everywhere, and f is unique, modulo h-null functions.

We now use Theorem 2.9 to prove some elementary inequalities involving the Young functions Φ, Ψ. At this point we extend Theorem 6.5 slightly, assuming that f is such that $\max\{m-1, \min(f(x),m)\}\, h$ is integrable for each integer m. We then say that f is h-*measurable*. The proof of the extended Theorem 6.5 still goes through, on modifying the sets W_j slightly and using a sequence (H_{1m}) instead of a fixed H_1. Thus if suitable Riemann sums are bounded we can take $r(x) = x^2$, $\Phi(x)$, $\Psi(x)$, for $x \ge 0$, and obtain integrability results. From the first, as $fg = \frac{1}{4}\{(f+g)^2 - (f-g)^2\}$, to obtain the integrability of fgh when f and g are h-measurable and the corresponding Riemann sums are bounded.

Theorem 12.2: *Let* $h(I,x) \geq 0$, $f(x) \geq 0$, $g(x) \geq 0$ *on* E^c, f *and* g *being* h-*measurable, and let* a,b *be positive constants. By Theorem* 2.9 (2.22),

$$(12.7) \qquad \int_E \Phi(af)dh + \int_E \Psi(bg)dh \geq ab \int_E fg\, dh,$$

assuming that the first two integrals exist, with equality if and only if

$$(12.8) \qquad \phi(af(x)) \leq bg(x) \leq \phi(af(x)+), \text{ h-almost everywhere.}$$

If for $a = a_0$, $b = b_0$, *the left side of* (12.7) *is* 1 *then*

$$(12.9) \qquad \int_E fgdh \leq a_0^{-1}b_0^{-1}.$$

Proof: (12.7) follows from the extended Theorem 6.5 and Theorem 2.9 (2.22) in the form

$$p(x) \equiv \Phi(af(x)) + \Psi(bg(x)) - abf(x)g(x) \geq 0.$$

If equality in (12.7), then for each integer $m > 0$, the set where $p(x) \geq 1/m$ has h-variation zero, so $p(x) = 0$ h-almost everywhere. When $p(x) = 0$, (12.8) follows. Conversely, (12.8) gives $p(x) = 0$, h-almost everywhere, and equality in (12.7). (12.9) is obvious, with (12.8) for equality.

Ex. 2.5 gives an important special case, with

$$\Phi(x) = x^p/p + \Psi(y) = y^q/q \quad (p > 1,\ 1/p + 1/q = 1),$$

$$(a^p/p)N(f,0)_p + (b^q/q)N(g,0)_q \geq abN(fg,0)_1.$$

If $N(f,0)_p > 0$, $N(g,0)_q > 0$, $a_0^{-1} = N(f,0)_p^{1/p}$, $b_0^{-1} = N(g,0)_q^{1/q}$, Hölder's inequality

(12.10) $N(fg,0)_1 \le N(f,0)_p{}^{1/p}N(g,0)_q{}^{1/q}$, $\int_E fgdh \le \left(\int_E f^p dh\right)^{1/p}\left(\int_E g^q dh\right)^{1/q}$,

follows, with equality when, h-almost everywhere,

(12.11) $a^{p-1}f^{p-1} = bg,\ a^p f^p = b^q g^q,\ f^p/g^q = \text{constant}.$

The inequality is still true when $N(f,0)_p = 0$, as $f = 0$ h-almost everywhere and $N(fg,0)_1 = 0$. Similarly when $N(g,0)_q = 0$. Replacing $1/p$ by t, $1/q$ by $1-t(0 < t < 1)$, f^n by f, g^q by g,

(12.12) $$\int_E f^t g^{1-t} dh \le \left(\int_E fdh\right)^t \left(\int_E gdh\right)^{1-t} \quad (f \ge 0,\ g \ge 0)$$

follows from (12.10), with equality when f/g is constant h-almost everywhere, the left side existing by Theorem 6.4 when the integrals on the right exist.

We denote by L^p the space of all h-measurable functions f on E^c for which $N(|f|,0)_p$ exists, writing $\|f\|_p = N(|f|,0)_p{}^{1/p}$. Note that throughout the theory we suppose f finite everywhere. When $p \to 1+$, then $q \to \infty$ and

(12.13) $$\left(\int_E g^q dh\right)^{1/q} \to \|g\|_\infty \equiv \text{ess sup } \{|g(x)| : x \in E^c\},$$

the supremum of all constants b for which the set X_b of x where $|g(x)| > b$, has $V(h;E;X_b) > 0$. We prove (12.13) by an easy inequality, and a second since the set of x with $|g(x)| > \|g\|_\infty$, has h-variation 0,

$$bV(h;E;X_b)^{1/q} \le \left(\int_E g^q\, dh\right)^{1/q} \le \|g\|_\infty V(h;E)^{1/q}.$$

We denote by L^∞ the space of all h-measurable g with $\|g\|_\infty$ finite, and (12.10) is true here with $p = 1$, "$q = \infty$". This follows since the set of all x with $|fg| > |f|\ \|g\|_\infty$, finite, has h-variation 0.

<u>Theorem 12.3</u>: (12.14) *(Minkowski's inequality). For the usual* $h \ge 0$, *if* f,g *are in* L^p, *so is* $f + g$, *and*

$$\|f+g\|_p \leq \|f\|_p + \|g\|_p \quad (1 \leq p \leq \infty).$$

(12.15) *Taking* $f \geq 0$, $g \geq 0$, *equality occurs when* $p = 1$. *When the finite* $p > 1$, *equality occurs if and only if for constants* $a \geq 0$, $b \geq 0$, $af(x) = bg(x)$ *h-almost everywhere.*

Proof: The case $p = 1$ is trivial. When the finite $p > 1$, $f + g$ is in L^p if f, g are. For if

$$|f(x)| \geq |g(x)|, \; |f(x)+g(x)|^p \leq (|f(x)| + |g(x)|)^p \leq (2|f(x)|)^p.$$

If

$$|f(x)| < |g(x)|, \; |f(x)+g(x)|^p \leq 2^p|g(x)|^p,$$

so

$$|f(x)+g(x)|^p \leq 2^p(|f(x)|^p + |g(x)|^p) \quad (\text{all } x).$$

Integrating, f+g is in L^p, and clearly we can replace f by $|f|$, g by $|g|$. Next, as $(p-1)q = p$, $|f|\,(|f| + |g|)^{p-1}h$ is integrable over E by Theorem 6.4 and (12.12). Similarly $|g|(|f| + |g|)^{p-1}h$ is integrable over E, and by Hölder's inequality, twice,

$$(12.16) \qquad \int_E (|f| + |g|)^p dh = \int_E |f|(|f|+|g|)^{p-1}dh + \int_E |g|(|f|+|g|)^{p-1}dh$$

$$\leq \left(\int_E |f|^p dh\right)^{1/p}\left(\int_E (|f|+|g|)^p dh\right)^{1/q} + \left(\int_E |g|^p dh\right)^{1/q}\left(\int_E (|f|+|g|)^p dh\right)^{1/q}.$$

(12.17) If $\int_E (|f| + |g|)^p dh = 0$,

then, h-almost everywhere, $|f| + |g| = 0$, $f = 0 = g$, and (12.14) is true. Otherwise, dividing by the qth root of the positive integral in (12.17), (12.16) gives (12.14). With (12.17) false and the finite $p > 1$, equality occurs in (12.14) if and only if both Hölder inequalities are equalities and for some constants a,b,

$|f|^p = a^q(|f|+|g|)^p$, $|g|^p = b^q(|f|+|g|)^p$ h-almost everywhere, and

$a'|f(x)| = b'|g(x)|$

for constants $a' \geq 0$, $b' \geq 0$.

The case "$p = \infty$" has to be considered separately. Here, $|f| \leq \|f\|_\infty$ except in a set X of h-variation 0, and $|g| \leq \|g\|_\infty$ except in a set Y of h-variation 0, and $X \cup Y$ has h-variation 0. Hence $|f+g| \leq \|f\|_\infty + \|g\|_\infty$ h-almost everywhere, and (12.14) follows in this case. Equality occurs if and only if, for each $\varepsilon > 0$, the sets X_ε and Y_ε where $|f| > \|f\|_\infty - \varepsilon$, $|g| > \|g\|_\infty - \varepsilon$, respectively, have $V(h;E;X_\varepsilon \cap Y_\varepsilon) > 0$.

Minkowski's inequality shows that the triangle inequality holds in L^p $(1 \leq p \leq \infty)$ if the "distance" from f to g is given by $\|g-f\|_p$, and this is a norm if f and g are identified when $f = g$ h-almost everywhere.

Having proved (12.10) when $1 \leq p \leq \infty$, we now show that the upper bound is attained.

Theorem 12.4: *For the usual* $h \geq 0$, *fixed* f *in* L^p, *and all* g *in* L^q *with* $\|g\|_q = 1$,

$$\sup \int_E fg\,dh = \sup \int_E |fg|\,dh = \|f\|_p \, \|g\|_q \quad (1 \leq p \leq \infty).$$

Proof: Without altering $\|g\|_q = 1$, we multiply g by $\operatorname{sgn}^{-1}(fg)$ where $\operatorname{sgn}(x) = x/|x|$ $(x \neq 0)$, $\operatorname{sgn}(0) = 1$, so that in the integrals, fg can be reaplced by $|fg|$, and we can assume $f \geq 0$, $g \geq 0$. When $p = 1$ we take $g(x) = 1$ (all x), so that $\|g\|_\infty = 1$ and the integral of $|fg|$ becomes $\|f\|_1$. When $1 < p < \infty$ we take $g(x) = |f(x)|^{p-1}/\|f\|_p^{p-1}$ so that

$$|g(x)|^q = |f(x)|^p/\|f\|_p^p, \; \|g\|_q = 1, \; |f(x)g(x)| = |f(x)|^p/\|f\|_p^{p-1}.$$

with integral $\|f\|_p$.

To prove this when "$p = \infty$", $q = 1$, we assume h VBG* *using a sequence* (X_j) *of* h*-measurable sets i.e. such that* $\chi(X_j;.)h$ *is integrable over* E. As f is h-measurable there are sets Y_m with union E^c such that $m-1 \leq f(x) \leq m$ for $x \in Y_m$, and such that $\chi(Y_m;.)h$ and $f\chi(Y_m;.)h$ are integrable over E for each m. Given $\varepsilon > 0$, the exact set X where $|f(x)| > \|f\|_\infty - \varepsilon$, has $0 < V(h;E;X) \leq \infty$ with $\chi(X;.)$ h-measurable. Using the integrability of fgh when f,g are h-measurable and suitable Riemann sums are bounded, we see that for $Z_{jm} = X_j \cap Y_m \cap X$, $\chi(Z_{jm};\cdot)h$ and $|f|\chi(Z_{jm};\cdot)h$ are integrable. As the union of the Z_{jm} is $E^c \cap X$ we choose j,m so that $Z = Z_{jm}$ has $0 < V(h;E;Z) < \infty$, and put

$$g(x) = V(h;E;Z)^{-1} \chi(Z;x), \quad \|g\|_1 = 1; \int_E |fg|dh =$$

$$\int_E |f| \chi(Z;.)dh/V(h;E;Z) \geq \|f\|_\infty - \varepsilon.$$

This proves Theorem 12.4 when "$p = \infty$", $q = 1$.

Returning to the general Young functions, we cannot always prove Theorem 12.2 (12.9) for constants a_0, b_0. For example, this cannot be done explicitly with the functions of Ex. 2.6. However, using Theorem 12.4's result as a definition we can proceed. Let L_Φ^* be the set of all real-valued h-measurable point functions f with $\Phi(|f|)$ integrable over E. Then L_Φ^* is not always linear. For in one dimension let $E = [0,1)$ and $h = m$ where $m([u,v),x) = v-u$. If $\Psi(y) = e^{-1}(e^y-1)$ (Ex. 2.6) and $f(x) = \log(x^{-\frac{1}{2}})$ then $\Psi(|f(x)|) = e^{-1}(x^{-\frac{1}{2}} - 1)$, integrable with respect to m. But $2.\log(x^{-\frac{1}{2}})$ does not lie in the set as $\Psi(|2f(x)|) = e^{-1}(x^{-1}-1)$, not integrable over [0,1) with respect to m.

To obtain linear sets of functions we define L_Φ to be the set of h-measurable point functions f for which $\|f\|_\Phi$ is finite, where for all h-measurable g with

$$\int_E \Psi(|g|)dh \leq 1, \quad \|f\|_\Phi = \sup \int_E |fg|dk,$$

the definition following the result of Theorem 12.4.

<u>Theorem 12.5</u> (12.19: *For the usual* $k \geq 0$, *the space* L_Φ *is linear, and if* f_1, f_2 *are in* L_Φ,

$$\|f_1 + f_2\|_\Phi \leq \|f_1\|_\Phi + \|f_2\|_\Phi .$$

(12.20) *If* $f \in L^*_\Phi$ *then* $\|f\|_\Phi \le \int_E \Phi(|f|)dh + 1$ *and* L_Φ *contains* L^*_Φ.

(12.21) *In the definition of* $\|f\|_\Phi$, $\int_E fg dh$ *can replace* $\int_E |fg| dh$.

(12.22) *Let* $h \geq 0$ *with variation not identically* 0, *be* VBG* *using a sequence* (X_j) *of* h*-measurable sets, and let* ϕ *be not identically* 0. *Then* $\|f\|_\Phi = 0$ *if and only if* $f = 0$ h*-almost everywhere.*

<u>Proof</u>: The linearity is clear, while if f_1, f_2 are in L_Φ and

$$\int_E \Psi(|g|)dh \le 1, \text{ then } \int_E |(f_1+f_2)g|dh \le \int_E |f_1 g|dh + \int_E |f_2 g|dh \le \|f_1\|_\Phi + \|f_2\|_\Phi .$$

Hence (12.19). As f,g are h-measurable, so are $\Phi(|f|)$, $\Psi(|g|)$, and we can consider the integrals below. Using Young's inequality, Theorem 2.9 (2.22),

$$\int_E |fg|dh \le \int_E \Phi(|f|)dh + \int_E \Psi(|g|)dh.$$

In the definition of $\|f\|_\Phi$ the last integral is not greater than 1, so that if the second integral is finite, $f \in L^*_\Phi$, then $f \in L_\Phi$ with (12.20). In the first integral of (12.21), replacing g by $g.\text{sgn}^{-1}(fg)$ we have the second integral. When $\|f\|_\Phi = 0$ with ϕ and the variation of h not identically 0, and h VBG* using h-measurable sets (X_j), we amalgamate every X_j of h-variation 0 with the first X_j of the sequence with positive h-variation, and so assume that $0 < V(h;E;X_j) < \infty$ $(j = 1,2,\ldots)$. As ϕ is not identically 0, there are points $r > 0$, $s > 0$ with

$$s = \phi(r),\ \psi(s) \le r,\ \psi(v) \le r (v \le s),\ \Psi(v) \le rv\ (v \le s),\ \lim_{v\to 0+} \Psi(v) = 0.$$

Of course the last result follows from (5.44) as soon as ψ is finite, but we need to prove ψ finite first, and the continuity is then a short step. Thus, for each positive integer j there are a number $p_j > 0$ and a function g_j on E^c such that

$$\Psi(p_j) \le V(h;E;X_j)^{-1}, \; g_j(x) = p_j \chi(X_j;x), \; \int_E \Psi(|g_j|)dh$$

$$= \int_E \Psi(p_j)\chi(X_j;x)dh = \Psi(p_j)V(h;E;X_j) \le 1, \; \|f\|_\Phi = 0,$$

$$0 = \int_E |fg_j|dh = \int_E |f|p_j \chi(X_j;x)dh, \; |f| \; \chi(X_j;.) = 0$$

h-almost everywhere. Hence $f = 0$ h-almost everywhere in X_j $(j = 1,2,\dots)$ and so h-almost everywhere in E^c, the union of the X_j. The converse is easy.

Theorem 12.6: (12.23) *If* $h \ge 0$, $f \in L_\Phi$, *but* $\|f\|_\Phi \ne 0$, *then* $f/\|f\|_\Phi$ *lies in* L_Φ^*.

(12.24) *If also there is a constant* $M > 0$ *such that* $\Phi(2x) \le M\Phi(x)$ $(x \ge x_0 \ge 0)$, *then* $L_\Phi = L_\Phi^*$.

(12.25) *(Hölder's inequality for Orlicz spaces). If* $f \in L_\Phi$, $g \in L_\Psi$, *and* f,g *are* h*-measurable, then* fgh *and* $|fg|h$ *are integrable and*

$$\left|\int_E fgdh\right| \le \int_E |fg|dh \le \|f\|_\Phi \, \|g\|_\Psi .$$

Proof: First we prove Ex. 2.4, to have

(12.26) $\Psi(qv) \ge q\Psi(v)$ $(q \ge 1, \; v \ge 0)$.

Being obvious for $q = 1$ or $v = 0$, we assume that $q > 1$, $v > 0$

$$\Psi(qv) - \Psi(v) = \int_{[v,qv)} \psi(x)dm \ge (q-1)v\psi(v) = (q-1)\int_{[0,v)} \psi(v)dm \ge$$

$$(q-1)\int_{[0,v)} \psi(x)dm = (q-1)\Psi(v).$$

If $f \in L_\Phi$, $\rho(g) \equiv \int_E \Psi(|g|)dh \le 1$, then $\int_E |fg|dh \le \|f\|_\Phi$, from the definition of $\|f\|_\Phi$. If $1 < \rho(g) < \infty$, (12.26) gives

$$\Psi(|g|) \le \rho(g)\Psi(|g|/\rho(g)),$$

$$\int_E \Psi(|g|/\rho(g))dh \le \int_E \Psi(|g|)dh/\rho(g) = 1,$$

$$\int_E |fg/\rho(g)|dh \le \|f\|_\Phi, \int_E |fg|dh \le \|f\|_\Phi \rho(g),$$

and so for all $g \in L_\Phi^*$.

$$\int_E |fg|dh \le \|f\|_\Phi \rho^*(g), \rho^*(g) \equiv \max(\rho(g),1).$$

As f is h-measurable, so are $|f|/\|f\|_\Phi$ and $\Phi(|f|/\|f\|_\Phi)$. Hence the integral of the latter exists if and only if suitable Riemann sums are bounded. If $g = \phi(|f|/\|f\|_\Phi)$ then g is h-measurable and we have equality in Young's inequality (Theorem 2.9 (2.22), (2.23)) at each point. As all the Riemann sums involved have non-negative terms, the following integrals exist and we have

$$\rho^*(g) \ge \int_E |(f/\|f\|_\Phi)g|dh = \int_E \Phi(|f|/\|f\|_\Phi)dh + \rho(g).$$

If $\rho(g) = \rho^*(g)$ then $\int_E \Phi(|f|/\|f\|_\Phi)dh = 0.$

If $\rho(g) < \rho^*(g)$ then $\rho^*(g) = 1$, $\int_E \Phi(|f|/\|f\|_\Phi)dh \le 1.$

Thus in either case (12.23) is true.

For (12.24) we have $L_\Phi^* \subseteq L_\Phi$. If $f \in L_\Phi$ and $\|f\|_\Phi \ne 0$ then $f/\|f\|_\Phi \in L_\Phi^*$. For some p, $\|f\|_\Phi \le 2^p$. Hence $f \in L_\Phi^*$ when $u_0 = 0$ since

$$\Phi(|f|) \le M^p\Phi(|f|/2^p) \le M^p \Phi(|f|/\|f\|_\Phi), \int_E \Phi(|f|)dh \le$$

$$M^p \int_E \Phi(|f|/\|f\|_\Phi)dh \le M^p.$$

If $u_0 > 0$ we write $f = f_1 + f_2$ where $f_1 = f$ on the set in E^c where $|f|/\|f\|_\Phi < u_0$, and $f_1 = 0$ elsewhere. Then $\Phi(|f_1|)$ is bounded by

$\Phi(u_0 \|f\|_\Phi)$ and is h-measurable, and so is integrable with respect to h. The integrability of $\Phi(|f_2|)h$ is proved as for $u_0 = 0$, and $\Phi(|f|) = \Phi(|f_1|) + \Phi(|f_2|)$, giving (12.24).

For (12.25), as f,g are h-measurable, fgh and $|fg|h$ are integrable. For if $\|g\|_\Psi = 0$, g = 0 h-almost everywhere and the result is trivial. Otherwise, replacing f, Φ by g, Ψ, in the proof of (12.23), $\rho(g/\|g\|_\Psi) \leq 1$, so that by definition

$$\int_E |fg|\,dh = \int_E |f(g/\|g\|_\Psi)\,dh.\ \|g\|_\Psi \leq \|f\|_\Phi\ \|g\|_\Psi .$$

In the statement (12.24) we can replace 2 by any number N > 1. Even so, the condition on Φ is very restrictive, and is not satisfied by the Ψ of Ex. 2.6.

At the beginning of the section, Theorem 12.1 proves the completeness of L^p spaces for all finite p > 0. We can now extend this to L_Φ spaces.

Theorem 12.7: *Let* $h \geq 0$ *be* VBG* *using a sequence* (X_j) *of mutually disjoint* h-*measurable sets. Then* L_Φ *is complete, i.e. if* $f_j \in L_\Phi$ $(j = 1,2,\ldots)$ *and* $\lim_{r,s\to\infty} \|f_r - f_s\|_\Phi = 0$, *there is an* $f \in L_\Phi$, *uniquely determined* h-*almost everywhere, such that* $\lim_{r\to\infty} \|f_r - f\|_\Phi = 0$.

Proof: Given $\varepsilon > 0$, there is an $N = N(\varepsilon)$ such that for all g with $0 < \rho(g) \leq 1$,

$$(12.27)\quad \int_E |f_r - f_s|.|g|\,dh \leq \varepsilon\ (r,s \geq N(\varepsilon)).$$

If $\phi = 0$ everywhere, so is Φ, L_Φ contains all h-measurable functions, and we have only to prove that the limit function of a sequence of h-measurable functions is h-measurable. See Section 16. So we proceed as in the proof of (12.22), obtaining $\Psi(x) \to 0$ $(x \to 0+)$ and assuming that $0 < V(h;E;X_j) < \infty$ $(j = 1,2,\ldots)$. Thus there is a constant $p_j > 0$ such that

$$\Psi(p_j) \leq V(h;E;X_j)^{-1}.\ \text{ If } g_j = p_j\chi(X_j;.),\ \int_E \Psi(|g_j|)dh = \Psi(p_j)V(h;E;X_j) \leq 1.$$

Using Theorem 12.1 on (12.27) with $g = g_j$, there are a point function f depending on j and a subsequence $(r_j(s))$ of the positive integers such that

$$\lim_{s\to\infty} f_{r_j(s)}(x) = f(x) \tag{12.28}$$

h-almost everywhere in X_j (g_j being zero elsewhere). By the proof of Theorem 12.1 we can arrange that for $j = 2,3,\ldots$, $(r_j(s))$ is a subsequence of $(r_{j-1}(s))$, so that for $r_j = r_j(j)$, r_k is in $(r_j(s))$ for each $k \geqq j$, and so, h-almost everywhere in E^c,

$$\lim_{j\to\infty} f_{r_j}(x) = f(x),$$

where by the usual mathematical economy of symbols, f(x) here is the f(x) in (12.28) when $x \in X_j$. Applying Fatou's lemma to (12.27) with $s = r_j$, uniformly in g with $\rho(g) \leq 1$,

$$\int_E |f_r - f|, |g| dh \leq \varepsilon \ (r \geq N(\varepsilon)), \quad \|f_r - f\|_\Phi \to 0 \text{ as } r \to \infty.$$

Again by Fatou's lemma, we have uniqueness as usual, and hence the theorem. Ex. 12.5 shows the need in the next theorem to have $f_j \in L_\Phi$, $g_j \in L_\Psi$, with bounded sequences of norms.

<u>Theorem 12.8</u>: *Let* $h \geq 0$ *on* E^c *with* $(f_j) \subseteq L_\Phi$, $(g_j) \subseteq L_\Psi$, *and let* f, g *be the respective limit of* (f_j), (g_j) *using* Φ, Ψ, *respectively* (*i.e.* $\|f_j - f\|_\Phi \to 0$, $\|g - g_j\|_\Psi \to 0$). *Also, for some constant* $M > 0$ *let* $\|f_j\|_\Phi \leq M$, $\|g_j\|_\Psi \leq M$ $(j = 1,2,\ldots)$. *Then* $f \in L_\Phi$, $g \in L_\Psi$, *and* $f_j g_j h$ *and* fgh *are integrable and*

$$\int_E f_j g_j dh \to \int_E fgdh.$$

<u>Proof</u>: As in the proof of Theorem 12.1, a subsequence $(j(r))$ of the positive integers satisfies $f_{j(r)}(x) \to f(x)$ h-almost everywhere. Then $(g_{j(r)}(x))$ is convergent using Ψ, and so for a subsequence $(k(r))$ of $(j(r))$, $g_{k(r)}(x) \to g(x)$ h-almost everywhere. Thus also $f_{k(r)}(x) \to f(x)$

h-almost everywhere. Thus if g is an h-measurable function with $\rho(g) \leq 1$, and using Fatou's lemma,

$$\int_E fgdh = \int_E \liminf_{r \to \infty} f_{k(r)}gdh \leq \liminf_{r \to \infty} \int_E f_{k(r)}gdh =$$

$$\liminf_{r \to \infty} \|f_{k(r)}\|_\Phi \leq M.$$

Hence $f \in L_\Phi$. Similarly $g \in L_\Psi$ on interchanging the roles of f, f_j and g, g_j. Hence $\rho(g/\|g\|_\Psi) \leq 1$ and so $f(g/\|g\|_\Psi)h$ is integrable. Similarly f_jg_jh, f_jgh, fg_jh are all integrable for each j. Finally, using Theorem 12.6 (12.25), Hölder's inequality here,

$$\left|\int_E f_jg_jdh - \int_E fgdh\right| = \left|\int_E \{(f_j-f)g + f(g_j-g) +\right.$$

$$\left.(f_j-f)(g_j-g)\}\, dh\right| \leq \|f_j-f\|_\Phi \quad \|g\|_\Psi$$

$$+ \|f\|_\Phi \|g_j-g\|_\Psi + \|f_j-f\|_\Phi \|g_j-g\|_\Psi \to 0 \text{ as } j \to \infty.$$

Note that if $f \in L_\Phi$ then by Theorem 12.5 (12.19),

$$\|f_j\|_\Phi \leq \|f_j - f\|_\Phi + \|f\|_\Phi ,$$

and the sequence $(\|f_j\|_\Phi)$ is bounded, and similarly for g_j. Thus the extra condition is necessary as well as being sufficient.

Ex. 12.1. A sequence $(f_j(x))$ can converge in mean without being pointwise convergent anywhere in E^c. For in one dimension put $f_j(x) = 0$ for $0 \leq x \leq 1$, except for two sloping sides of a triangle, the third side being on the x-axis. As $j \to \infty$ let the area of the triangle tend to 0 while the triangle moves repeatedly from 0 to 1 with height tending to infinity.

Ex. 12.2. If a sequence $(f_j(x))$ is pointwise convergent and convergent in mean, then the limits are the same h-almost everywhere. (Use Fatou's lemma.)

Ex. 12.3. For one dimension with $h(I,x)$ the length m of I, let a_j be real with $\sum_{j=1}^{\infty} a_j^2$ convergent. Then $f_j(x) = \sum_{s=1}^{j} a_s \sin sx$ is convergent in mean with index $p = 2$, to a function $f(x)$ over $E^c = [-\pi,\pi]$. For if $E = [-\pi,\pi)$,

$$\int_E \sin sx \sin tx \, dm = \begin{cases} 0 & (s \neq t) \\ \pi & (s = t) \end{cases} \quad \text{and} \quad \int_E (f_t - f_s)^2 \, dm =$$

$$\pi \sum_{j=s+1}^{t} a_j^2 \quad (s < t).$$

Ex. 12.4. Show that the sequences $(s_j(x))$, $(t_j(x))$ converge in mean with index 2 over $E = [-\pi,\pi)$ where $s_j(x) = \sum_{u=2}^{j} u^{-1} (\log u) \sin ux$, $t_j(x) = \sum_{u=2}^{j} (u \log u)^{-1} \sin ux$. If f,g are the respective functions to which s_j, t_j are convergent in mean, find

$$\int_E f(x)g(x)dm$$

(New University of Ulster, 1977, M 313)

Ex. 12.5. Let (f_j) be convergent in mean to f with index p and let g be any h-measurable function. Then $(f_j + g)$ is convergent in mean to $f + g$ with index p.

CHAPTER 4

LIMIT THEOREMS FOR MORE GENERAL CONVERGENCE, WITH CONTINUITY

13 Basic Theorems

We can replace the integer variable j of Chapter 3 by a variable y that takes more than a sequence of values, and lies in a directed set Y, and then we often run into difficulties. Some of the difficulties can be seen when y is a real variable in $[0,\infty)$ and we take the limit as $y \to \infty$. Putting $y = (w-a)^{-1}$ and $(a-w)^{-1}$, we can deal with limits $w \to a+$ and $w \to a-$, respectively, and so can consider the continuity of an integral with respect to a parameter w at the point $w = a$. At this stage it is thus enough to give the tests of Sections 8, 9, 10, replacing $j \to \infty$ by $y \to \infty$, and a general Y replacing $j \to \infty$ in Section 11.

Theorem 13.1: *For* E *an elementary set,* $k(I,x) \geq 0$ *in* E^c*, and* $f(x,y)$ *real-valued and monotone increasing in* y *for* y *in* $[0,\infty)$*, let* $f(x,y)k(I,x)$ *be integrable over* E *to* $H(E,y)$*. If* $H(E,y)$ *is bounded above for* $y \geq 0$*, with supremum* $H(E)$*, then* $f(x,y)$ *tends to a limit* $f(x)$ *as* $y \to \infty$*, except in a set* X *of* x *with* $V(k;E;X) = 0$ *(i.e. the limit exists* k*-almost everywhere) and* $f(x)\chi(X;x)k(I,x)$ *is integrable over* E *to* $H(E)$*.*

Proof: Restricting y to the integers j, by Theorems 8.1, 8.2, $f(x,j)$ tends to a limit $f(x)$ as $j \to \infty$, except in such a set X, and $f(x)\chi(X;x)k(I,x)$ is integrable over E, to the supremum of $H(E,j)$ in j, which is $H(E)$ by monotonicity in y. Similarly $f(x,y) \to f(x)$ as $y \to \infty$, except in X, and $\lim_{y\to\infty} H(E,y)$ exists equal to $H(E)$.

Theorem 13.2: *For* E *an elementary set,* $k(I,x) \geq 0$ *on* E^c*, and* $f(x,y)$ *real-valued for* $x \in E^c$*, let* $f(x,y)k(I,x)$ *be integrable to* $H(E,y)$ *over* E *for each* $y \geq 0$*. If there are a real number* B *and a positive function* δ *on* E^c *such that for all* δ*-fine divisions* $\mathcal{D}$ *of* E *and all choices of functions* $y(x)$*,*

$$(13.1) \qquad (\mathcal{D}) \sum f(x,y(x))k(I,x) \geq B,$$

if $\lim_{y\to\infty} f(x,y)$ *exists equal to* $f(x)$, *finite,* k-*almost everywhere, and if* $\liminf_{y\to\infty} H(E,y)$ *is finite then* fk *is integrable over* E *and*

$$(13.2)\qquad \int_E f\,dk \le \liminf_{y\to\infty} \int_E f(x,y)\,dk = \liminf_{y\to\infty} H(E,y).$$

If (13.1) *is replaced by*

$$(13.3)\qquad (\mathcal{D})\,\Sigma\, f(x,y(x))k(I,x) \le B$$

for the same δ, $\mathcal{D}$, $y(x)$, *with* $f(x,y)$ *tending to a finite limit* $f(x)$ k-*almost everywhere and* $\limsup_{y\to\infty} H(E,y)$ *finite, then* fk *is integrable over* E *and*

$$(13.4)\qquad \int_E f\,dk \ge \limsup_{y\to\infty} \int_E f(x,y)\,dk = \limsup_{y\to\infty} H(E,y).$$

If (13.1), (13.3) *are replaced by*

$$(13.5)\qquad B \le (\mathcal{D})\,\Sigma\, f(x,y(x))k(I,x) \le C$$

for some real numbers $B < C$ *and for all choices of functions* $y(x)$, *and if the other conditions hold, then* fk *is integrable over* E *to* $H(E) \equiv \lim_{y\to\infty} H(E,y)$, *a finite limit, i.e.*

$$(13.6)\qquad \int_E \lim_{y\to\infty} f(x,y)\,dk = \lim_{y\to\infty} \int_E f(x,y)\,dk.$$

Proof: There is a sequence (y_j) of y tending to $+\infty$, such that

$$\lim_{j\to\infty} \int_E f(x,y_j)\,dk = \liminf_{y\to\infty} \int_E f(x,y)\,dk.$$

Applying Theorem 9.1 (9.2) to $f_j(x) = f(x,y_j)$, we have (13.2) since, k-almost everywhere,

$$\liminf_{j\to\infty} f_j(x) = \lim_{y\to\infty} f(x,y).$$

Similarly for (13.4), and then (13.6) follows.

If in the proof of (13.2) we drop the condition that $\lim_{y \to \infty} f(x,y)$ exists (finite) k-almost everywhere, then Theorem 9.1 (9.2) gives the existence of the integral of $\liminf_{j \to \infty} f(x,y_j)$, which can be put in place of the first integral in (13.2). As this lower limit is not less than $\liminf_{y \to \infty} f(x,y)$, we can multiply by k and integrate it over E, to replace the first integral in (13.2), provided that the integral exists. But Ex. 13.1 shows that we cannot take the integral's existence for granted. By Theorem 13.1 the integral does exist if $\inf_{0 \le y \le m} f(x,y).k$ is integrable over E for each $m > 0$. A similar discussion can be made of (13.4), replacing every inf by sup, while (13.6) need not be altered.

Similar changes can be made of the majorized (dominated) convergence test.

Theorem 13.3: *Let* y *be a real-valued parameter, let* E *be an elementary set, let* $k(I,x) \ge 0$, *and let* $f(x,y)$, $p(x)$, $q(x)$ *be point functions, all defined on* E^c. *Let* $f(.,y)k$, pk, qk *all be integrable in* E.

If $p \le f(.,y)$ *for all* y, *with finite*

(13.7) $$\liminf_{y \to \infty} \int_E f(.,y)dk, \text{ and } \inf_{a \le y \le m} f(x,y).k \quad (\text{all } a < m)$$

integrable, then $\liminf_{y \to \infty} f(.,y)k$ *is integrable over* E *and*

(13.8) $$\int_E \liminf_{y \to \infty} f(.,y)dk \le \liminf_{y \to \infty} \int_E f(.,y)dk.$$

On the other hand, if $f(.,y) \le q$ *for all* y, *with finite*

(13.9) $$\limsup_{y \to \infty} \int_E f(.,y)dk, \text{ and } \sup_{a \le y \le m} f(x,y).k \quad (\text{all } a < m)$$

integrable, then $\limsup_{y \to \infty} f(.,y)k$ *is integrable over* E *and*

(13.10) $\limsup_{y \to \infty} \int_E f(.,y)dk \leq \int_E \limsup_{y \to \infty} f(.,y)dk.$

(13.11) *If* $f(x) \equiv \lim_{y \to \infty} f(x,y)$ *exists (finite)* k-*almost everywhere, and if* $p \leq f(.,y) \leq q$ *(all* y*), then* fk *is integrable and* (13.6) *holds. Here,* (13.7), (13.9) *are not needed.*

Proof: Use Theorem 13.2.

We now extend the theorems on controlled convergence in one dimension.

Lemma 13.4: (13.12) *For each* $y \geq 0$ *let* $f(.,y)$ *be a real or complex valued point function on* $[a,b]$ *that converges almost everywhere in* $[a,b]$ *to a finite* f *as* $y \to \infty$, $f(.,y)m$ *being integrable to* $F(y;I)$ *on all intervals* $I \subseteq [a,b)$.

(13.13) *Let* $F(y;.)$ *be* AC* *in* $[a,b)$ *uniformly in* y, *so that the* $F(y;.)$ *are absolute integrals.*

(13.14) *Then* $V(F(y;.) - F(u;.); [a,b)) \to 0$ *as* $y,u \to \infty$.

(13.15) *Further, if* $F(y;.) \to F$ *as* $y \to \infty$ *then* $V(F-F(u;.); [a,b)) \to 0$ *as* $u \to \infty$.

Proof: As in the proof of Lemma 10.1 we can assume that $f(.,y)$ converges everywhere. For integers p and given $\varepsilon > 0$, let Y_p be the set of x where

(13.16) $|f(x,y) - f(x)| < \varepsilon$ (all $y \geq p$).

Then we again prove that, uniformly in y,

(13.17) $V(F(y;.);[a,b);\sim Y_p) < \varepsilon(p \geq p_0(\varepsilon))$, $V(F(y;.)-F(u;.);[a,b)) \leq 2\varepsilon(b-a+1)$

$(y > u \geq p_0(\varepsilon))$,

giving (13.14), (13.15).

Lemma 13.5: *Assuming* (13.12) *with*

(13.18) *the* $F(y;[a,x))$ *converging uniformly on* $[a,b]$ *to* $F([a,x))(a < x \le b)$,

(13.19) *the* $F(y;.)$ *are* $AC^*(X)$ *uniformly in* y *where* X *is a closed subset of* $[a,b]$,

(13.20) *then* $V(F(y;.) - F(u;.); [a,b);X) \to 0$, $V(F-F(u;.);[a,b);X) \to 0$ $(y < u, y \to \infty)$.

Theorem 13.6: *If* (13.12) *and* (13.18) *hold and*

(13.21) *the functions* $F(y;.)$ *are* ACG^* *uniformly in* y,

(13.22) *then* fm *is integrable on* $[a,b)$ *and*

$$\int_{[a,b)} f(.,y)dm \to \int_{[a,b)} fdm \quad (y \to \infty).$$

The proofs of Lemma 13.5 and Theorem 13.6 parallel the proofs of Lemma 10.2 and Theorem 10.3.

Next we examine the necessary and sufficient conditions, using a directed set Y of parameters y, and the limit process $\lim_Y$ of Y.

Theorem 13.7: (13.23) *Let* E *be an elementary set, let the real-valued function* $k(I,x)$ *of bounded variation be defined in* E^c, *and for each* $y \in Y$ *let* $f(.,y):E^c \to R$ *with* $f(.,y)k$ *integrable in* E, *such that* $\lim_Y f(x,y) = f(x)$ *for each fixed* $x \in E^c$. *A necessary and sufficient condition for the integrability of* fk *over* E, *is that*

(13.24) $(\mathcal{D}) \sum f(x,y(x))k(I,x) \in S$

for some compact set S *of arbitrarily small diameter, a positive function* δ *on* E^c, *a function* $M:E^c \to Y$, *all functions* $y(x):E^c \to Y$ *satisfying* $y(x) \geq M(x)$ *everywhere on* E^c *in the direction of* Y, *and all* δ*-fine divisions* $\mathcal{D}$ *of* E.

Theorem 13.8: *Given* (13.23) *and the integrability of* fk *over* E, *a necessary and sufficient condition for*

$$\lim_{Y} \int_E f(.,y)dk = \int_E \lim_{Y} f(.,y)dk$$

is that for some compact set T *of arbitrarily small diameter with* $S \cap T$ *not empty, and a* $J \in Y$, *a positive function* $\delta(.,y)$ *on* E^c *for each* $y \in Y$ *with* $y \geq J$, *and all* $\delta(.,y)$*-fine divisions* $\mathcal{D}$ *of* E,

$$(13.25) \quad (\mathcal{D}) \sum f(x,y)k(I,x) \in T \quad (y \geq J)$$

See the proofs of Theorems 11.1, 11.2.

These results are specialized to differentiation in Section 15, and to integration in Section 18.

Ex. 13.1. In one dimension let X be a set in $[0,1)$ for which its characteristic function or indicator $\chi(X;x)$ is not integrable over $[0,1)$ with respect to $m(I,x)$, defined as the length of I. For $0 \leq x \leq 1$, $0 \leq y < 1$, and $q = 1,2,\ldots$, let

$$f(x,y) = \begin{cases} 0 & (x = y \in X) \\ 1 & (\text{otherwise}) \end{cases} \qquad f(x,y+q) = f(x,y).$$

Then f is bounded in $[0,1] \times [0,+\infty)$ and is integrable with respect to m in $[0,1)$ for each fixed y, while

$$\liminf_{y \to \infty} f(x,y) = \inf_{0 \leq y \leq 1} f(x,y) = 1 - \chi(X;x)$$

is not integrable.

A construction of such an X, using the multiplicative axiom, or Zorn's lemma, of set theory, is as follows. Let $Z(x)$ be the set of all z in $0 \leqq z < 1$ with $z-x$ rational. Let X be a set in $[0,1)$ such that for each x in $[0,1)$, $X \cap Z(x)$ contains exactly one point. It is the existence of X that needs the multiplicative axiom. If (r_j) is the sequence of mutually distinct rationals in $[0,1)$ let X_j be the set of all $x + r_j$, $x + r_j - 1$ that lie in $[0,1)$, with $x \in X$. Then X_j splits up into two parts, each congruent to a part in a corresponding split of X. Hence if the second integral exists,

$$\int_{[0,1)} \chi(X_j;x)dm = \int_{[0,1)} \chi(X;x)dm,\ k\int_{[0,1)} \chi(X;x)dm =$$

$$\int_{[0.1)} \chi(\bigcup_{j=1}^{k} X_j;x)dm \leqq 1,\quad \int_{[0,1)} \chi(X;x)dm = 0,$$

$$\int_{[0,1)} \chi(\bigcup_{j=1}^{\infty} X_j;x)dm = 0.$$

e.g. use the monotone convergence theorem. But $\bigcup_{j=1}^{\infty} X_j = [0,1)$, giving a contradiction. Hence $\chi(X;x)m$ cannot be integrated over $[0,1)$.

14 Fatou's Lemma And The Avoidance Of Nonmeasurable Functions

This section deals with the problem of obtaining Fatou-type results when sequences of functions depending on an integer parameter j, are replaced by functions depending on a parameter y lying in a uncountable directed set Y having a limit process $\lim_{y \in Y}$. We have to avoid non-measurable functions such as turn up in Ex. 13.1.

Theorem 14.1: (14.1) *Let* E *be an elementary set, let* $h(I,x) \geq 0$ *in* E^c, *and for each* $y \in Y$ *let* $f(.,y) : E^c \to R$ *be such that* $f(.,y)h$ *is integrable over* E.

(14.2) *If for some* $u \in Y$ *there is a compact set* S *depending on* u, *such that for each sequence* $(y_j) \subseteq Y$ *with every* $y_j \geq u$, *some positive function* δ *on* E^c *and depending on* (y_j) *and* u, *each* δ*-fine division* $\mathcal{D}$ *of* E, *and each integer-valued function* $j(x)$ *on* E^c,

$$(\mathcal{D}) \sum f(x,y_{j(x)})h(I,x) \in S, \text{ then } g(x,(y_j)) \equiv \inf_{j\geq 1} f(x,y_j)$$

exists (finite) h*-almost everywhere and is integrable over* E, *say to* $G((y_j))$. *As* (y_j) *varies in* Y *with every* $y_j \geq u$, *then* $G((y_j))$ *has an infimum, say* $J(u)$, *there is a* $(z_j) \subseteq Y$ *with every* $z_j \geq u$ *and* $G((z_j)) = J(u)$, *and there is a function* $k(u;.) : E^c \to R$, *independent of* (y_j) *and integrable over* E *to* $J(u)$, *such that* $g(x,(y_j)) \geq k(u,x)$ h*-almost everywhere, with* $g(x,(y_j)) = k(u;x)$ h*-almost everywhere when* $G((y_j)) = J(u)$. *Conversely, if the* $G((y_j))$ *exist and have an infimum, then* (14.2) *holds. Further,*

$$(14.3) \quad J(u) \leq \inf_{y\geq u} \int_E f(x,y)dh.$$

We write $k(u;x)$ as $\inf^*_{y\geq u} f(x,y)$, h-measurable. By the construction, h-almost everywhere,

$$k(u;x) \geq \inf_{y\geq u} f(x,y),$$

but the last lower limit need not be h-measurable, see Ex. 13.1.

<u>Proof</u>: From (14.2) for each positive integer m,

$$(14.4) \quad (\mathcal{D}) \sum \min_{1\leq j\leq m} f(x,y_j)h(I,x) \in S, \text{ and } \min_{1\leq j\leq m} f(x,y_j)h(I,x)$$

is integrable by Theorem 6.3. Theorem 8.2 (strong monotone convergence) with (14.4) gives the integrability of $g(x,(y_j))h$ to a value in S. Thus $G((y_j))$ has an infimum $J(u)$ in S, and so finite. Hence for each $q = 1,2,\ldots$ there is a sequence (y_{qj}) with $G((y_{qj})) < J(u) + 1/q$. For (z_j) the sequence $y_{11}, y_{12}, y_{21}, y_{13}, y_{22}, y_{31}, y_{14}, \ldots$, each (y_{qj}) is a subsequence, $G((z_j)) \leq G((y_{qj})) < J(u) + 1/q$ $(q = 1,2,\ldots)$, $G((z_j)) = J(u)$.

Writing $k(u;x) = g(x,(z_j))$, we show that it is unique modulo values on sets of h-variation 0. For if (w_j) satisfies $G((w_j)) = J(u)$ let (t_j) be the sequence $z_1,w_1,z_2,w_2,z_3,\ldots$. Then

$$G((t_j)) \le G((z_j)) = J(u),\ G((t_j)) = J(u),\ g(x,(t_j)) \le g(x,(z_j)),$$

$$\int_E (g(x,(z_j)) - g(x,(t_j))dh$$

$$= J(u)-J(u) = 0,\ g(x,(t_j)) = g(x,(z_j))$$

h-almost everywhere; so, similarly, $g(x,(t_j)) = g(x,(w_j))$ h-almost everywhere, and $g(x,(w_j)) = g(x,(z_j)) \equiv k(u;x)$ h-almost everywhere. Conversely, if $J(u)$ is finite, (14.4) follows from

$$(14.5)\quad J(u) \le G(y_j)) \le \int_E f(x,y_j)dh\ (j = 1,2,\ldots),$$

for some positive function δ on E^c and depending on (y_j) on u, and then (14.2). Further, taking any particular $y \in Y$ in $y \ge u$ as a member of (y_j), (14.5) gives (14.3).

For $f(.,y)$ monotone increasing in y, $k(u;x) = f(x,u)$ will suffice. But if $f(.,y)$ is monotone decreasing in y we need the full construction except when Y has a linear order so that in each finite subset of Y there is a greatest y. This has relevance in the proof of the first part of the next theorem, as $k(u;x)$ is monotone increasing in each sequence of u, h-almost everywhere. We need the limit as u rises in Y, obtained using $\sup^*_{u\in Y}$.

Theorem 14.2: (14.6) *Let the conditions of Theorem* 14.1 *hold for all* $u \in Y$. *Writing* $\sup^*_{u\in Y} k(u;x)$ *as* $\liminf^*_{y\in Y} f(x,y)$, *we have a Fatou-type result,*

$$(14.7)\quad \int_E \liminf^*_{y\in Y} f(x,y)dh \le \lim_{u\in Y}\ \inf_{y\ge u} \int_E f(x,y)dh$$

in the sense that if the right side is finite then the left side exists with the inequality. Similarly,

$$(14.8)\qquad \int_E \limsup_{y\in Y}{}^{*} f(x,y)dh \geq \lim_{u\in Y}\sup_{y\geq u} \int_E f(x,y)dh,$$

(14.9) *and if* $\liminf_{y\in Y}{}^{*} f(x,y) = \limsup_{y\ Y}{}^{*} f(x,y)$ h-*almost everywhere,*

and written as $\lim_{y\in Y}{}^{*} f(x,y)$,

$$(14.10)\qquad \int_E \lim_{y\in Y}{}^{*} f(x,y)dh = \lim_{y\in Y} \int_E f(x,y)dh.$$

Naturally, $\sup_{u\in Y}{}^{*} r(x,u) = -\inf_{u\in Y}{}^{*} (-r(x,u))$, $\limsup_{y\in Y} r(x,y) =$

$$-\liminf_{y\in Y}{}^{*} (-r(x,y)).$$

The proof of this theorem is now straightforward.

Ex. 14.1. Let Y be the collection of all finite sets y of real numbers. For y,u ∈ Y let y ≤ u have the meaning y ⊆ u, and let

$$f(x,y) = \sup_{z\in Y} g(x,z),\quad g(x,z) = \begin{cases} 0 & (x \neq z) \\ 1 & (x = z) \end{cases}$$

Then f = 0 = g except at a finite number of points, and so for m([u,v),x) = v-u,

$$\lim_{y\ Y}{}^{*} f(x,y) = 0$$

almost everywhere. But

$$\lim_{y\in Y} f(x,y) = \sup_{y\in Y} f(x,y) = 1$$

everywhere.

CHAPTER 5

DIFFERENTIATION, MEASURABILITY AND INNER VARIATION

15. Differentiation Of Integrals

First we consider the differentiation of the integral with respect to a parameter,

$$(15.1)\quad \frac{d}{dy}\int_E f(x,y)dk \equiv \lim_{h\to 0}\int_E \frac{f(x,y+h) - f(x,y)}{h}\, dk$$

$$= \int_E \lim_{h\to 0} \frac{f(x,y+h) - f(x,y)}{h}\, dk \equiv \int_E \frac{\partial f(x,y)}{\partial y}\, dk.$$

The case in one dimension of $k = m$ where $m([u,v),x) = v-u$, aroused great interest early in the history of the calculus as it was realised that if the process could be legitimised, many difficult integrals could be computed. As a result, C. Jordan in 1891 or earlier, requested the following, quoted by C. de la Vallée Poussin (1892).

> 'Give a rigorous theory of differentiation under the integral sign of definite integrals, with precise conditions which limit Leibnitz's rule, principally for unbounded regions of integration or unbounded functions, and particularly many celebrated definite integrals.'

After a century, we can give necessary and sufficient conditions for which (15.1) holds. Here, h is real, so that Section 13 can be used and we can apply various tests in particular cases, even when the elementary set E is changed to intervals $[a,+\infty)$, $(-\infty,b)$, $(-\infty,+\infty)$, or infinite volumed bricks in higher dimensions, provided that we use the integral of Section 5 after Theorem 5.14, or the analogue in higher dimensions. The tests assume $h > 0$. Similar tests hold when $h < 0$. For notation we use

$$R(f;x,y,h) \equiv \{f(x,y+h)-f(x,y)\}/h,\ R(H;E,y,h) \equiv \{H(E,y+h)-H(E,y)\}/h.$$

<u>Theorem 15.1</u>: (15.2) *Let* E *be an elementary set,* $k(I,x) \geq 0$ *in* E^c, *and* $f(x,z)$ *real-valued for* $x \in E^c$, *with* $f(.z)k$ *integrable over* E *to* $H(E,z)$,

for each z *in a neighbourhood of* y, *so that* R(f;x,y,h)k *is integrable over* E *to* R(H;E,y,h) (h > 0).

(15.3) *If* R(f;x,y,h) *is monotone in* h > 0 *and the monotone* R(H;E,y,h) *is bounded as* $h \to 0+$, *then* R(f;x,y,h) *tends to a finite limit* k*-almost everywhere as* $h \to 0+$, *that we can write as* $(\partial f/\partial y)_{y+}$, *and* (15.1) *is true on the right at* y.

Instead of monotonicity we can use bounded Riemann sums.

Theorem 15.2: *Let* (15.2) *hold. If there are a real number* A *and a positive function* δ *on* E^c *such that for all* δ*-fine divisions* $\mathcal{D}$ *of* E *and all positive functions* h *on* E^c,

(15.4) $(\mathcal{D})\sum R(f;x,y,h(x))k(I,x) \geq A$,

(15.5) *if* $\lim_{h\to 0+} R(f;x,y,h) \equiv (\partial f/\partial y)_{y+}$ *exists (finite)* k*-almost everywhere,*

(15.6) *and if* $\liminf_{h\to 0+} R(H;E,y,h)$ *is finite,*

then $(\partial f/\partial y)_{y+}k$ *is integrable over* E *and*

(15.7) $$\int_E (\partial f/\partial y)_{y+}dk \leq \liminf_{h\to 0+} \int_E R(f;x,y,h)dk = \liminf_{h\to 0+} R(H;E,y,h).$$

If with (15.5) *and the same* δ, $\mathcal{D}$, h(x), (15.4) *is replaced by*

(15.8) $(\mathcal{D})\sum R(f;x,y,h(x))k(I,x) \leq B$,

(15.9) *with* $\limsup_{h\to 0+} R(H;E,y,h)$ *finite,*

then again $(\partial f/\partial y)_{y+}k$ *is integrable in* E, *with*

(15.10) $$\int_E (\partial f/\partial y)_{y+}dk \geq \limsup_{h\to 0+} \int_E R(f;x,y,h)dk = \limsup_{h\to 0+} R(H;E,y,h).$$

If (15.4), (15.8) *hold for some real numbers* $A < B$ *and all choices of positive functions* h, *with* (15.5), *then* (15.6), (15.9) *hold with* (15.1) *for right-hand derivatives.*

Next, Theorem 13.3 gives a majorized (dominated) convergence test.

Theorem 15.3: *Assuming* (15.2), *let* p,q *be real-valued functions with* $p \leq q$ *on* E^c *and* pk, qk *integrable on* E. *Then* (15.7) *follows from* (15.5), (15.6), *and*

$$(15.11) \quad R(f;x,y,h) \geq p(x) \quad (x \in E^c)$$

while (15.10) *follows from* (15.5), (15.9), *and*

$$(15.12) \quad R(f;x,y,h) \leq q(x) \quad (x \in E^c).$$

If (15.2), (15.5), (15.11), (15.12) *hold then* (15.1) *follows for right-hand derivatives.*

The corresponding theorem for controlled convergence in one dimension is as follows.

Theorem 15.4: *Assuming* (15.2) *with* $E = [a,b)$, *let*

(15.13) $R(H;[a,x),y,h)$ *converge to* $(\partial H([a,x),y)/\partial y)_{y+}$ *as* $h \to 0+$, *uniformly for* x *in* $(a,b]$.

(15.14) *let the* $R(H;[a,x),y,h)$ *be* ACG* *uniformly in* $h > 0$.

Then (15.1) *is true for right-hand derivatives.*

Next follow necessary and sufficient conditions for (15.1) with right-hand derivatives.

Theorem 15.5: *Assuming* (15.2), (15.5), *a necessary and sufficient condition for the integrability of* $(\partial f/\partial y)_{y+}k$ *over* E, *is that for some compact set* S *of arbitrarily small diameter, two positive functions* δ, M *on* E^c, *all positive functions* h *on* E^c *satisfying* $h \leq M$ *everywhere on* E^c, *and all δ-fine divisions $\mathcal{D}$ of* E,

(15.15) $(\mathcal{D})\,\Sigma R(f;x,y,h(x))k(I,x) \in S.$

Given this integrability, a necessary and sufficient condition for (15.1) *with right-hand derivatives, is that for some compact set* T *of arbitrarily small diameter with* $S \cap T$ *not empty, a* $J > 0$*, a positive function* $\delta(\cdot,h)$ *on* E^c *for each* h *in* $0 < h \leq J$*, and all* $\delta(\cdot,h)$*-fine divisions $\mathcal{D}$ of* E,

(15.16) $(\mathcal{D})\,\Sigma R(f;x,y,h)k(I,x) \in T \ (0 < h \leq J).$

The proofs of these theorems follow the proofs of corresponding theorems in Section 13, with y changed to h and f(x,y) to R(f;x,y,h). There is a similar set of theorems for left-hand derivatives. In order that two-sided derivatives occur, we take the limits in (15.5) and the analogue to be equal k-almost everywhere.

In classical theory the differentiation of integrals relative to bricks in one dimension is appreciably easier than in n dimensions ($n > 1$), as in the latter we have to avoid needle-like bricks in order that Vitali's covering theorem can be used. Our theory gives the size of exceptional sets, showing that they are of inner variation zero, and then in special cases that they are of variation zero. Let E be an elementary set and δ a positive function on the closure E^c. A definition from Section 7 is that a set Q of brick-point pairs is a δ-fine partial division of E if Q is a subset of a δ-fine division of E. A set C of brick-point pairs (I,x) is called an *inner cover* of a set X in R^n, if for each positive function δ on E^c and each $x \in X$, there is a δ-fine pair $(I,x) \in C$ with $I \subseteq E$. The set X *has inner* k*-variation* 0 if for each $\varepsilon > 0$ there is an inner cover C of X such that for each δ-fine partial division Q of E with $Q \subseteq C$, and k(I,x) defined for E^c,

(15.17) $(Q)\ \Sigma|k(I,x)| < \varepsilon.$

Sierpinski's lemma when $n = 1$, and Vitali's theorem when $n > 1$, are used to find in some sense a maximal Q out of C, to prove that the variation of k over X is 0.

Let $p(I,x)$, $q(I,x)$ be defined for E^c. If, for a particular $x \in E^c$ a number d exists with the property that, given $\varepsilon > 0$, there is a $\delta > 0$ such that for all δ-fine (I,x) with $I \subseteq E$, then

(15.18) $\quad |p(I,x) - d.q(I,x)| \leq \varepsilon|q(I,x)|.$

If another number d^* ($\neq d$) satisfies the condition then

$$|d-d^*|\ |q| = |p-d^*q + dq-p| \leq 2\ \varepsilon\ |q|.$$

For $\varepsilon = |d-d^*|/4$, $q = 0$ follows. Otherwise, if d exists it is unique, and we write d as $d(p,q;E;x)$. It is independent of E if a sphere centre x, lies in E. We call d the *derivative of* p *relative to* q *and* E. As the set X of points x where $q(I,x) = 0$ for some positive function δ on E^c and all δ-fine (I,x), clearly has $V(q;E;X) = 0$, X can be disregarded.

Theorem 15.6: (15.19) *For* E *an elementary set and* $f(x)$, $k(I,x)$ *defined for* E^c, *let* fk *be integrable to* H *on the partial sets of* E. *Then* $d(H,k;E;x) = f(x)$ *except at points of a set* $X \subseteq E^c$ *with inner* k*-variation* 0. *Also the set* X_1 *of all* x *where* $d(H,k;E;x)$ *exists but is not* $f(x)$, *has* k*-variation* 0.

(15.20) *If* (X_j) *is a sequence of sets each with inner* k*-variation* 0, *the union of the sets also has inner* k*-variation* 0.

Proof: There is a positive function ε on X such that for each positive function δ on E^c and each $x \in X$,

(15.21) $\quad |H(I) - f(x)k(I,x)| \geq \varepsilon(x)|k(I,x)|$

for at least one δ-fine (I,x). Then X is the union of sets X_j on which $2^j\varepsilon(x) > 1$. By Theorem 5.3, for each positive integer j there is a positive function δ_j on E^c such that for each δ_j-fine division $\mathcal{D}_j$ of E,

$$(15.22) \quad (\mathcal{D}_j)\ \Sigma|H(I) - f(x)k(I,x)| \le \varepsilon 4^{-j}.$$

Taking $\delta = \delta_j$ in $X_j \setminus X_{j-1}$ with X_0 empty, δ arbitrary in $\setminus X$, and $\mathcal{D}$ a δ-fine division of E, then by (15.21), (15.22),

$$(\mathcal{D}_j)\Sigma|k(I,x)|\chi(X_j;x) \le \varepsilon 2^{-j},\ (\mathcal{D})\Sigma|k(I,x)|\chi(X;x) \le \varepsilon,$$

and X has inner k-variation 0. The proof of (15.20) follows similarly.

To show that X_1 has k-variation 0, in (15.21) we can take $\varepsilon(x) > |d-f(x)|/2$, the inequality being true for all δ-fine (I,x) and some positive δ on E^c. Then Theorem 5.7 (5.22) with $2/|d-f(x)|$ replacing f(x), shows that $V(k;E;X_1) = 0$.

Theorem 15.7: *In Theorem* 15.6 *let* n = 1 *with* k, fk, |fk| *integrable in an elementary set* E *and* k ≥ 0. *If* K *is the integral of* k *then*

$$\lim_{v\to u+} \frac{1}{K([u,v))} \int_{[u,v)} |f(x)-f(u)|dk = 0 \quad ([u,v) \subseteq E)$$

except for x *in a set of inner* k*-variation* 0. *Similarly for* v → u-.

Proof: For any constant c, the integrability of k, fk and |fk|, imply the integrability of (f-c)k; and |f-c|k, = max(f-c,c-f)k when f-c is real-valued, using Theorem 6.3; or by Ex. 6.4 when f-c is complex-valued. Thus by Theorem 15.6 we have

$$\lim_{v\to u+} \frac{1}{K([u,v))} \int_{[u,v)} |f(x)-c|dk = |f(u)-c| \quad ([u,v) \subseteq E)$$

except for u in a set X(c) of inner k-variation 0. Taking $c = \pm p.2^{-q} \pm ir.2^{-s}$ when f is complex-valued, and $\pm p.2^{-q}$ when f is real-valued (p,q,r,s = 0,1,2,...), the union X of the countable number of sets X(c) for the special c, is also of inner k-variation 0, by Theorem 15.6 (15.20). If b is any constant and c is one of the special numbers with

$$|c-b| \le \varepsilon,\ \left|\int_{[u,v)} |f(x)-b|dk - \int_{[u,v)} |f(x)-c|dk\right| \le \int_{[u,v)} \big||f(x-b|-|f(x)-c|\big|d$$

$\leq |c-b|K([u,v)) \leq \varepsilon K([u,v))$.

For $u \in \setminus X$, some positive function δ^* on E^c, and $|v-u| \leq \delta^*$ (c), Theorem 15.6 gives

$$\int_{[u,v)} |f(x)-c|\,dk \leq K([u,v)) \quad \{|f(u)-c| + \varepsilon\} \leq K([u,v))\{|f(u)-b| + 2\varepsilon\},$$

$$\int_{[u,v)} |f(x)-b|\,dk \leq K([u,v)) \quad \{|f(u)-b| + 3\varepsilon\}.$$

The arbitrary constant b can depend on u. Taking $b = f(u)$, the theorem follows.

Theorem 15.8: *If* $p(I,x)$, $q(I,x)$ *be defined for* E^c *with* q *arbitrary and* p VB* *or* VBG* *in a set* X, *relative to* E, *then* $p(I,x)/q(I,x)$ $(I \subseteq E)$ *is either of the form* 0/0 *or is bounded as* $\delta \to 0$ *for* δ*-fine* (I,x) $(x \in X)$, *except for* $x \in X_1$, X_1 *being of inner* q*-variation* 0. *If* $V(p;E;X) = 0$ *then* $d(p,q;E;x) = 0$ *in* X, *except for a set of inner* q*-variation* 0.

Proof: By Theorem 15.6 (15.20) it is enough to take p VB*. If for a constant M, $|p| > M|q|$ in an inner cover C, and if $Q \subseteq C$ is a δ-fine partial division of E then for small δ,

$$M(Q)\, \Sigma\, |q| \leq (Q)\, \Sigma\, |p| \leq (\mathcal{D})\, \Sigma\, |p| \leq V(p;\delta;E;X) < \infty.$$

Where the ratio is unbounded, $M \to \infty$ and the inner q-variation is 0. If $V(p;E;X) = 0$ then $V(p;\delta;E;X)$ is as small as we please, whatever the $M > 0$, and by Theorem 15.6 (15.20) the set of all points with a fixed $M > 0$, has inner q-variation 0. Hence the second part.

Theorem 15.9: *Let* p,q,r,s *be brick-point functions defined for* E^c, *let* p,r *be variationally equivalent, and let* q,s *be variationally equivalent, both relative to* X *and* E. *Then, for* $x \in X$, either

$$d(p,q;E;x) = d(r,s;E;x)$$

or else neither derivative exists, except for x *in a set of inner* q-*variation* 0.

Proof: If $d = d(p,q;E;x)$ exists for a particular $x \in X \cap E^c$, then omitting the (I,x),

$$|r-ds| \leq |r-p| + |p-dq| + |d|\ \ |q-s|,\ V(r-p;E;X) = 0 = V(q-s;E;X),$$

$$|r-p| \leq \varepsilon|q|,\ |q-s| \leq \varepsilon\ |q|$$

by Theorem 15.8, except for x in a set of inner q-variation 0. As

$$|p-dq| \leq \varepsilon|q|,\quad |r-ds| \leq (\varepsilon + \varepsilon + |d|\varepsilon)|q|,$$

when (I,x) is δ-fine and δ is small enough, except for x in a set of inner q-variation 0, and so of inner s-variation 0, as the exceptional set is in X. Hence the result as we can interchange p,q with r,s.

This could have been used to prove the first part of Theorem 15.6.

Two geometrical results now follow.

Theorem 15.10: *In one dimension let* E *be an elementary set, with* C *an inner cover of* X *formed of* (I,x) *with* $I = [x,u) \subseteq E$ *for some* $u > x$, *and let* $k(I,x)$ *be continuous in* E^c *with finite* $V(k;E;X)$. *Then, given* $\varepsilon > 0$, *there is a finite number of mutually disjoint intervals from* C *with union* Y, *such that*

(15.25) $\quad V(k;E;X \cap Y) > V(k;E;X) - \varepsilon.$

Let $|k|$ *also be integrable on* E *with the inner* k-*variation of* X *as* 0. *Then the* k-*variation of* X *is* 0.

Proof: Clearly we can assume that E is an interval $L = [a,b)$ and then add up the results at the end. Let X_j be the set of $x \in X$ for which there is an $[x,u)$ from C with $u-x > 1/j$. Then $X = \lim_{j\to\infty} X_j$ so that by Theorem 5.11 we can choose j so that

(15.24) $V(k;L;X_j) > V(k;L;X) - \frac{1}{2}\varepsilon$.

(15.25) Put $a_1 = \inf X_j \geq a$, $b_1 = \sup X_j \leq b$, $c = b_1 - a_1$, $\eta = \frac{1}{2}\varepsilon/(jc+1)$.

By Ex. 5.10, $V(k;I)$ is continuous as the ends of I vary, and if the interval $J \subseteq I$, then

(15.26) $V(k;L;X \cap J) \leq V(k;I)$,

the left side being continuous in J, so that there is a point $x_1 \in X_j$ such that either $x_1 = a_1$ or, by (15.26),

(15.27) $V(k;L;X \cap [a_1,x_1)) < \eta$.

Let $[x_1,u_1) \in C$ with $u_1 - x_1 > 1/j$. If points of X_j lie to the right of u_1 let a_2 be their infimum. Then either $a_2 = x_2 \in X_j$ or, by (15.26) and the continuity of h, there is an $x_2 \in X_j$ with $V(k;L;X \cap [a_2,x_2)) < \eta$. Let $[x_2,u_2) \in C$ with $u_2 - x_2 > 1/j$. And so on. If in J such steps we are at b_1, then by (15.24), (15.25), (15.27) and similar results,

$$X_j \subseteq Y^c \cup Z^c,\ Y \equiv \bigcup_{j=1}^{J} [x_j,u_j),\ Z \equiv \bigcup_{j=1}^{J} [a_j,x_j),$$

$$\begin{aligned} V(k;L;X) - \tfrac{1}{2}\varepsilon < V(k;L;X_j) &\leq V(k;L;X \cap Y^c) + \sum_{j=1}^{J} V(k;L;[a_j,x_j]) \\ &\leq V(k;L;X \cap Y) + J\eta, \end{aligned}$$

by continuity of k. Also

$$(J-1)/j < c,\ J < cj+1,\ J\eta < \tfrac{1}{2}\varepsilon,\ V(k;L;X) < V(k;L;X \cap Y) + \varepsilon,$$

proving (15.23). Now let $|k|$ also be integrable in E, so that the integral is $V(k;E)$ by the Theorem 5.4 (5.13). If the inner k-variation is 0 let the inner cover C give (15.17), the members being δ-fine for some positive function δ on E^c that satisfies Theorem 5.3 (5.6) with h replaced by $|k|$, H by $V(k;.)$, and 8ε by ε. Then by continuity of k, and

(15.23),

$$V(k;L;X \cap Y) = V(k;Y;X) \leq V(k;Y) = \sum_{j=1}^{J} V(k;[x_j,u_j))$$

$$< \sum_{j=1}^{J} |k([x_j,u_j),x_j)| + \varepsilon < 2\varepsilon,\ V(k;L;X) < 3\varepsilon.$$

As $\varepsilon > 0$ is arbitrary, $V(k;L;X) = 0$.

The inner covers considered in Theorem 15.10 could be called right-hand inner covers. As we are in one dimension every inner cover of a set X can be split up into two inner covers, a right-hand inner cover C_1 of a set $X_1 \subseteq X$, and a left-hand inner cover C_2 of a set $X_2 \subseteq X$ where $X_1 \cup X_2 = X$ and the members of C_2 are of the form $([u,x),x)$ for some $u < x$. The proof of Theorem 15.10 also applies to left-hand inner covers, with obvious changes, so that if X is of inner k-variation 0 then

$$0 \leq V(k;\acute{E};X) \leq V(k;E;X_1) + V(k;E;X_2) = 0.$$

For dimension $n > 1$ the next covering theorem needs bricks that are near to cubes. Let I be the Cartesian product of n intervals $[a_j,b_j)(j = 1,\ldots,n)$ and write

$$\mu(I) = \prod_{j=1}^{n} (b_j-a_j),$$

the product of the lengths of the edges, *i.e. the* n-*dimensional volume*. This has to be used instead of a more general $k(I,x)$, at least until Theorem 15.11 is improved. The coefficient of regularity $r(I) = \sup \mu(I)/\mu(J)$ for all cubes $J \supseteq I$. If $d = \max_{1\leq j\leq n} (b_j-a_j)$ then $r(I) = \mu(I)d^{-n}$. If there is a number $\alpha > 0$ such that all the $(I,x) \in C$ have $r(I) \geq \alpha$, we call C an α-*inner cover*, and this is used to define when a set X has α-*inner variation* 0. In the definition of a derivative we also insert the condition $r(I) \geq \alpha$. Theorem 15.7 can be given for $n > 1$ by replacing $K([u,v))$ by $k(I,u)$ with u fixed, though to use Theorem 15.11 we have to take $k = \mu$.

Theorem 15.11: *For fixed* $\alpha > 0$ *and the set* X, *let* C *be an* α*-inner cover of* $X \cap E^c$, *and let* F *be the collection of closures of bricks that occur in the elements of* C. *Then there is a finite or countable sequence* (I_j^c) *of mutually disjoint closed bricks of* F *such that*

$$(15.28) \qquad V(\mu;E;X \setminus \bigcup_{j=1}^{\infty} I_j^c) = 0.$$

Proof: We define I_j^c by induction. Let I_1^c be an arbitrary closed brick of F. When the mutually disjoint closed bricks $I_1^c,\dots,I_{j-1}^c$ have been defined, if their union U includes all $X \cap E^c$ we need go no further. Otherwise there is an $x \in X \cap E^c$ that does not lie in U, which is closed. Thus there is a sphere $S(x,\delta(x))$ that is disjoint from U. By definition C contains a δ-fine (I,x), so that there are $I^c \in F$ that have no points in common with U. Let d_j be the supremum of diameters of such I^c. Then d_j is finite as the original elementary set E is of finite diameter. Let I_j^c be any one of those I^c with diameter greater than $\frac{1}{2}d_j$. Then the inductive definition proceeds. If we obtain an infinite sequence (I_j^c) we put

$$Y = X \cap E^c \setminus \bigcup_{j=1}^{\infty} I_j^c.$$

(15.29) *Suppose that* $V(\mu;E;Y) > 0$.

As for each j, $r(I_j) \geq \alpha > 0$, we can associate with each I_j^c a closed cube K_j such that

$$(15.30) \qquad I_j \subseteq K_j, \ \mu(I_j) \geq \alpha\mu(K_j).$$

Let L_j be the brick that is a cube with the same centre as K_j but with diameter $(4n+1)$ times as great, where of course n is the dimension. The following series is convergent since by (15.30),

$$(15.31) \qquad \sum_{j=1}^{\infty} \mu(L_j) = (4n+1)^n \sum_{j=1}^{\infty} \mu(K_j) \leq (4n+1)^n \alpha^{-1} \sum_{j=1}^{\infty} \mu(I_j^c)$$
$$\leq (4n+1)^n \alpha^{-1} \mu(E).$$

Now $\mu > 0$ is finitely additive and continuous. Hence by Theorem 5.6, Exx. 5.8, 5.9 and (15.29), there is an integer N such that

$$V(\mu;E;\bigcup_{j=n+1}^{\infty} L_j) \leq \sum_{j=N+1}^{\infty} V(\mu;E;L_j) = \sum_{j=N+1}^{\infty} V(\mu;L_j)$$

$$= \sum_{j=N+1}^{\infty} \mu(L_j) < V(\mu;E;Y).$$

It follows that there is an $x \in Y$ not belonging to any $L_j (j > N)$. By construction of Y,

$$x \in \setminus \bigcup_{j=1}^{\infty} I_j^c.$$

(15.32) *Hence there are a sphere* $S(x,\delta_1(x))$ *with no points in common with the closed*

$$\bigcup_{j=1}^{N} I_j^c,$$

and a closed brick $J^c \in F$ *with* $J \subseteq E$ *and* δ_1*-fine* (J,x).

Now the diameter of J, $\mathrm{diam}(J) > 0$, while (15.31) is a convergent series so that $\mu(K_j) \to 0$ and K_j is a cube. Hence we cannot have, for all j,

$$\mathrm{diam}(J) \leq d_j \leq 2\mathrm{diam}(I_{j+1}) \leq 2\mathrm{diam}(K_{j+1}).$$

So, for some j, $d_j < \mathrm{diam}(J)$. By definition of d_j, J must have points in common with

$$\bigcup_{p=N+1}^{j} I_p^c.$$

Let s be the smallest j for which $J \cap I_j^c$ is not empty. Then J is disjoint from I_j^c for $j = 1,2,\ldots,s-1$, and

(15.33) $\mathrm{diam}(J) \leq d_{s-1}$.

By (15.32), $s > N$, so that by definition of x, $x \notin L_s$. Thus J contains points outside L_s and points of $I_s{}^c \subseteq K_s{}^c$. But $\text{diam}(L_s) = (4n+1)\text{diam}(K_s)$, so

$$\text{diam}(J) > 4n(\text{half edge of } K_s) = 2n(\text{edge of } K_s) > 2\text{diam}(K_s) \geq 2\text{diam}(I_s) > d_{s-1},$$

contradicting (15.33). Thus from (15.29) we get a contradiction, so (15.29) is false and the theorem is true.

Theorem 15.12: *Let* X *be the union of sets* X_α $(\alpha = 1/p,\ p = 1,2,\ldots)$ *for which* X_α *is of* α*-inner variation* 0, *relative to the elementary set* E. *Then* $V(\mu;E;X) = 0$.

Proof: By Theorem 5.6 we need only prove the result for a constant $\alpha > 0$. Given $\varepsilon > 0$, there is an α-inner cover C of X such that every partial division $Q \subseteq C$ of E satisfies

$$(Q)\ \Sigma\mu\,(I) < \varepsilon.$$

Using this with Theorem 5.6, 15.11 (15.28), and Exx. 5.8 (for $n > 1$), 5.9, as $\mu > 0$ is finitely additive and continuous,

$$V(\mu;E;X) \leq V(\mu;E;X \setminus \bigcup_{j=1}^{\infty} I_j) + \sum_{j=1}^{\infty} V(\mu;E;X \cap I_j)$$

$$\leq \sum_{j=1}^{\infty} V(\mu;E;I_j) = \sum_{j=1}^{\infty} V(\mu;I_j) = \sum_{j=1}^{\infty} \mu(I_j).$$

Being true for each $\varepsilon > 0$, $V(\mu;E;X) = 0$.

The proof of Theorem 15.11 is so rigidly dependent on the use of μ that one needs to find what $k(I,x)$ can replace μ. For the definition and use of inner variation are for quite general $k(I,x)$. Exx. 15.4, 15.5 are offered as a start of this research, and one can combine the ideas in the two examples.

Ex. 15.1. Let $a < b$, $y \geq 0$, and consider

$$f(y) \equiv \int_{[a,b)} \frac{1}{1+ye^x}\, dm \quad (m([u,v),x) = v-u).$$

Find $f'(y)$ and hence or otherwise evaluate the integral.

(New University of Ulster 1974, M212)

Ex. 15.2. By finding $f'(y)$ show that

$$f(y) \equiv \int_{[0,\pi)} \sec x \log_e(1 + y\cos x)dm = \pi \arcsin y \ (|y| < 1)$$

(New University of Ulster 1975, M211)

Ex. 15.3. If in n dimensions

$$h(I) = \begin{cases} \mu(I) & (\mu(I) \text{ rational}) \\ 0 & (\mu(I) \text{ irrational}) \end{cases}$$

then any brick has inner h-variation 0 while $V(h;I) = \mu(I)$, h is not integrable over any brick, and h is not variationally equivalent to $V(h;\cdot)$ in any brick.

Ex. 15.4 Let $g_j(x)$ be continuous and strictly increasing in $(-\infty,+\infty)$ for $1 \leq j \leq n$. Show that in Theorem 15.11, μ can be replaced by μ_g, the product of differences $g_j(b_j) - g_j(a_j)$, provided that a suitable change is made in the definition of the coefficient of regularity.

Ex. 15.5. If $0 < f(x) \leq h(x)$ on E^c with

$$f(x)\,\mu(I) \leq |k(I,x)| \leq h(x)\mu(I),$$

then a set X of α-inner k-variation 0 is of α-inner variation 0, so giving $V(\mu;E;X) = 0$ and hence $V(k;E;X) = 0$.

Ex. 15.6. (J.J. McGrotty) In one dimension let $k([u,v),x) = \begin{cases} 1 & (u < 0 = v) \\ 0 & (\text{otherwise}) \end{cases}$. Then $k = |k|$ is integrable over $E = [-1,1)$ while each point x of E has an interval $[x,v)$ attached with $v > x$ and $k = 0$. Taking $X = E$, $V(k;E;X) = 1$ but $V(k;E;X \cap X_1) = 0$ in Theorem 15.10, showing that (15.23) is false here, and that the theorem needs a condition like the continuity of k.

Ex. 15.7. In one dimension let f,g be point functions for $I = [u,v)$ let $\Delta(f(I) = f(v)-f(u)$. If $V(\Delta f\Delta g;E) = 0$ show that f is continuous except in a set X with inner Δg-variation zero. If also g is continuous in E and VB* in X relative to E, then $V(\Delta g;E;X) = 0$.

Ex. 15.8. In one dimension we can have $v(\Delta f\Delta g;E) = 0$ and yet $\int fdg$ and $\int gdf$ do not exist. For take $g(x) = x$(all x) and $f(x) = -1(x\log_e x)$ $(0 < x \leq \frac{1}{2})$, $f(0) = 0$, with finite derivatives in $(0,\frac{1}{2})$ and $\Delta f\Delta g$ continuous at the ends. Thus

$$V(\Delta f\Delta g;[0.\tfrac{1}{2}) = 0,\ \int_{[\varepsilon,\frac{1}{2})} f(x)dg = \log_e\log_e(1/\varepsilon) - \log_e\log_e 2$$

which does not tend to a limit as $\varepsilon \to 0+$. By Theorem 5.12, $\int_{[0,\frac{1}{2})} g(x)df$ does not exist either.

Ex. 15.9. (a) Prove that, for all $[u,v)$ in an elementary set E,

$$\int_{[u,v)} fdg = f(v)g(v)-f(u)g(u)$$

if and only if $V(g^*\Delta f;E) = 0$, where g^* denotes that g is evaluated at the opposite end of the interval I from the x in (I,x).

(b) If in (a), X_j is the set of x in E^c for which $|g(x)| \geq 1/j$, then prove that f is continuous in X_j^c for each j, and the union X of the X_j' (derived sets) has inner Δf-variation 0.

(c) If in (b) f is continuous and VB* in E^c prove that $V(\Delta f;E;X) = 0$.

(d) If $[a,b] \subseteq X_j'$ for some j, prove that f is constant in $[a,b]$.

Hint: $f(x)\{g(x)-g(t)\} - \{f(x)g(x)-f(t)g(t)\} = g(t)\{f(t)-f(x)\}$, while in (d), X_j is everywhere dense in $[a,b]$. Now see Theorem 3.5.

Ex. 15.10. Let $|k|$ and $|k|\ \chi(X);.)$ be integrable in an elementary set E, where X is a set of x, and where we are in one dimension. If $a \leq x \leq b$, $a < b$, $[a,b) \subseteq E$, and $b-a \to 0$, then $V(k;[a,b);X)/V(k;[a,b)) \to \chi(X;x)$ k-almost everywhere.

In Lebesgue theory this is called the *density theorem*, and an x for which the ratio tends to 1, is called a *point of metric density* 1 *of* X. In higher dimensions we have a similar result using bricks with coefficient of regularity $\geq \alpha$ replacing $[a,b)$ and with $k = \mu$.

Ex. 15.11. Problem EFG 54 of the Bulletin of the Institute of Mathematics and its Applications (vol. 21, July/August 1985, p. 143) is as follows. A well-behaved non-negative function f satisfies $f(u) \to 0$ as $u \to -\infty$ and

$$\int_{(-\infty,u)} f(x)^p dm = \left(\int_{(-\infty,u)} f(x)dm\right)^p$$

for some constant $p > 0$, and $m([u,v),x) = v-u$. Find f.

If $p = 1$ we only have the integrability of f. If $p \neq 1$ we differentiate to obtain, almost everywhere,

$$\text{(a)}\quad f(u)^p = p\left(\int_{(-\infty,u)} f(x)dm\right)^{p-1} f(u) \qquad (u \geq 0).$$

Assuming $f(u) \neq 0$, then $f = g$ almost everywhere where (a) is true everywhere when f is replaced by g. Thus $f(u) = A \exp(au)$ where $A > 0$ is a constant and $a > 0$ satisfies $p = a^{p-1}$.

Clearly trouble occurs when $f(u) = 0$. Let X_1 be the set of x where $f(x) > 0$, assuming that f is not 0 almost everywhere. Using *Ex.* 15.10 let X be the set of points of X_1 that are of metric density 1 of X_1. Then $X_1 \setminus X$ has variation zero and

$$\text{(b)}\quad f(u) = \chi(X;u)a \int_{(-\infty,u)} f(x)dm$$

almost everywhere and $f = g$ almost everywhere, where g satisfies (b) everywhere and so is AC* and monotone increasing, and, almost everywhere,

$$g'(u) = \chi(X;u)ag(u), \quad d\log_e g(u)/du = a\chi(X;u).$$

Then for some $b \in X$ and for all $x > b$,

$$\log_e g(x) = aV(m;[b,x);X), \quad g(x) = A\exp(aV(m;[b,x);X)) \quad (x>b,\ x \in X),$$

where A is a positive constant. Similarly, for a positive constant A_1, we have

$$g(x) = A_1\exp(-aV(m;[x,b);X)) \qquad (x < b,\ x \in X).$$

As b is a point of metric density of X_1 and so of X, and by the continuity of g on X, we can let $x \to b$ to obtain $g(b) = A = A_1$. Thus f satisfies, almost everywhere,

$$F(x) = \begin{cases} \chi(X;x)A\exp(aV(m;[b,x);X)) & (x > b) \\ \chi(X;x)A\exp(-aV(m;[x,b);X)) & (x < b) \end{cases}$$

for constants $A > 0$ and $b \in X$, such that $f(x) \to 0$ as $x \to -\infty$ in the set of variation zero where the last formula does not hold. $A = 0$ covers the case where $f = 0$ almost everywhere.

16 Limits Of Step Functions

A function $g(x)$ is a step function in the elementary set E if there are a division $\mathcal{D}$, enumerated as $(I_j,x_j)(j = 1,2,\ldots,J)$ and constants g_j such that $g(x) = g_j$ in the interior of I_j $(j = 1,\ldots,J)$. As the μ-variation of the boundary of a brick is 0, we see that whatever finite values $g(x)$ takes on the boundaries of the I_j, $g(x)\mu(I)$ is integrable in E with

$$(16.1) \quad \int_E g(x)d\mu = \sum_{j=1}^{J} g_j \, \mu(I_j).$$

Theorem 16.1: *Let* $f(x)\mu(I)$ *be integrable to* $F(P)$ *over every partial set* F *of* E. *Then* f *is the limit almost everywhere of a sequence of step functions, the integral of each step function over* E *being equal to* $F(E)$.

Proof: We can assume that E is a brick I, which we divide into 2^{pn} equal bricks I_j $(j = 1,2,3,\ldots,2^{pn})$ by continued bisection with respect to each co-ordinate in turn, so that the coefficient of regularity $r(I_j) = r(I) > 0$. We put

$$s_p(x) = \int_{I_j} fd\mu/\mu(I_j) \quad (x \in I_j{}^o, \; j = 1,\ldots,2^{pn}).$$

$$(16.2) \quad \int_I s_p(x)d\mu = \sum_{j=1}^{2^{pn}} \left\{\int_{I_j} fd\mu/\mu(I_j)\right\}\mu(I_j) = \sum_{j=1}^{2^{pn}} \int_{I_j} fd\mu = \int_I fd\mu = F(I).$$

By Theorems 15.6, 15.11, F is differentiable to f except in a set X of variation 0, when the coefficients of regularity are bounded below away from 0. If $x \notin X$ and is not on any boundary then

$$s_p(x) = F(I_j)/\mu(I_j) \to f(x)$$

as required, and hence the theorem.

A further result, on the differentiation of finitely additive brick functions of bounded variation, follows in a similar way.

Theorem 16.2: *Let the real* $H(I)$ *be finitely additive in the bricks* $I \subseteq E$ and of bounded variation in the elementary set E. Then

(16.3) $f(x) = d(E;\mu;E;x)$ *exists almost everywhere in* E, *and*

(16.4) *if we put* $f(x) = 0$ *in the exceptional set then* f *and* $|f|$ *are integrable in* E *with*

$$\int_E |f|\,d\mu \leq V(H;E).$$

Proof: By Theorem 7.2 we can write H as the difference of two non-negative finitely additive brick functions

(16.5) $H(J) = \frac{1}{2}\{V(H;J) + H(J)\} - \frac{1}{2}\{V(H;J) - H(J)\}.$

Thus we can assume $H \geq 0$.
For numbers $M > 0$, $\alpha > 0$, suppose that a set X has the property that

(16.6) for some α-inner cover C of $X \cap E^c$, $H \geq M\mu$.

Given $\varepsilon > 0$, there is a positive function δ on E^c such that

(16.7) $V(H;\delta;E;X) < V(H;E;X) + \varepsilon.$

By Vitali's theorem (Theorem 15.11) there is a finite collection Q of disjoint bricks with union U, from C, such that

$$M\{V(\mu;E;X) - \varepsilon\} \leq M.V(\mu;E;X \cap U) \leq M.V(h;E;U) = M.(Q)\Sigma\mu(J)$$

$$\leq (Q)\Sigma H(J) \leq V(H;\delta;E;X) < V(H;E;X) + \varepsilon,$$

using (16.6), (16.7). As $\varepsilon \to 0+$ we have

(16.8) $V(H;E;X) \geq M.V(\mu;E;X).$

If we can take M arbitrarily large then $V(\mu;E;X) = 0$. Also, if (16.6) holds for C and if $H \leq N\mu$ for another α-inner cover of $X \cap E^c$, where $M > N$, then

$$M.V(\mu;E;X) \leq V(H;E;X) \leq N.V(\mu;E;X),\ (M-N)V(\mu;E;X) \leq 0,$$

(16.9) $V(\mu;E;X) = 0.$

Taking $M = (j+1)2^{-p}$, $N = j.2^{-p}$ $(j,p = 0,1,2,\ldots)$, each set X satisfies (16.9). Taking $\alpha = 1/q(q = 1,2,\ldots)$ and taking the countable union of all such sets with M,N,α, Theorem 5.6 shows that this union satisfies (16.9). In the complement of this union the finite derivative exists, proving (16.3).

We again take each brick I of E and divide I into 2^{pn} equal bricks $I_j(j = 1,2,3,\ldots,2^{pn})$ with $s_p(x) = H(I_j)/\mu(I_j)$ $(x \in I_j^{o}, j = 1,2,\ldots,2^{pn})$ and by finite additivity,

$$\int_I s_p(x)d\mu = \sum_{j=1}^{2^{pn}} (H(I_j)/\mu(I_j))\mu(I_j) = \sum_{j=1}^{2^{pn}} H(I_j) = H(I).$$

Remembering that $H \geq 0$ and that $s_p(x) \to f(x)$, the derivative, except on the boundaries and except in a set of variation 0, Fatou's lemma (say, Theorem 9.1 (9.2)) gives the existence of

$$\int_I f d\mu = \int_I \lim_{p\to\infty} s_p(x)d\mu \leq \liminf_{p\to\infty} \int_I s_p(x)d\mu = H(I).$$

For arbitrary H, $d(H;\mu;E;x)$ is a difference f_1-f_2 of two integrable non-negative functions, except in a set of variation 0, while

$$\int_I |f|d\mu \leq \int_I f_1 d\mu + \int_I f_2 d\mu \leq \tfrac{1}{2}\{V(H;I)+H(I)\} + \tfrac{1}{2}\{V(H;I)-H(I)\}$$

$$= V(H;I),$$

giving (16.4).

This section is tightly linked with Theorem 15.11 and so with μ. Any widening of Theorem 15.11 will result in better theorems here. There are wider theorems in one dimension.

<u>Theorem 16.3</u>: *In one dimension let* $k(I,x) \geq 0$, *defined for* E^c, *be continuous there, and let* $f(x)$ *be defined on* E^c. *Let* k *and* fk *be respectively integrable to* K *and* F *over the partial sets of the elementary set* E. *Then there is a sequence* (s_p) *of step functions such that* $\lim_{p\to\infty} s_p = f$ k-*almost everywhere in* E^c *and*

$$\int_E s_p dk = \int_H fdk.$$

Proof: $K \geq 0$ by Theorem 4.4 (4.5) and is continuous by (5.44). If $K([a,b)) = 0$, then as K is finitely additive (Theorem 5.1), $K([u,v)) = 0$ for all $[u,v) \subseteq [a,b)$, and then

$$\int_{[u,v)} fdk = 0$$

by Theorem 5.7, changing k into K. This change also occurs even if $K \neq 0$. We now follow the proof of Theorem 16.1 using the continuity of K and the definition

$$s_p(x) = F(I_j)/K(I_j) \quad (x \in I_j^{\,o},\ j = 1,2,3,\dots,2^p),$$

$$\int_I s_p(x)dK = \sum_{j=1}^{2^p} \{F(I_j)/K(I_j)\}K(I_j),$$

this being a Riemann-Stieltjes integral, and Theorem 15.10 replacing Theorem 15.12.

Theorem 16.4: *In one dimension, given an elementary set* E *let* $k(I,x) \geq 0$, *defined for* E^c, *be continuous there, and let* $h(I,x)$ *be defined for* E^c, *such that* h, $|h|$, k *all be integrable in* E. *Then* $f(x) \equiv d(h,k;E;x)$ *exists in* E^c *except for a set* X *of* x *of* k*-variation* 0. *If we put* $f = 0$ *in* X, *then* fk *and* $|f|k$ *are integrable in* E *with*

$$\int_E |f|dK \leq V(h;E).$$

Proof: Follow the proof of Theorem 16.2, making obvious changes.

Theorem 16.5: (16.10) *If* h *is* k-AC* (k-ACG*) *relative to* X *and* E, *where* E *is an elementary set,* $h(I,x)$ *and* $k(I,x)$ *are defined for* E^c, *and* k *is continuous in* E^c *and* VB* *in* X, *then* h *is* VB*(VBG*) *in* X.

(16.11) *In one dimension, if* $h(I,x), k(I,x)$ *are defined for the closure* E^c *of the elementary set* E*, and if* h *is integrable in* E *and* AC* *relative to* k*, where* $k \geq 0$ *is integrable (and so* VB*) *in* E *and continuous in* E^c*, then the finite derivative* $f \equiv d(h,k;E;.)$ *exists* k*-almost everywhere.*

(16.12) *In* (16.11), *putting* $f(x) = 0$ *where* $d(h,k;E;x)$ *does not exist, then*

$$\int_I dh = \int_I fdk.$$

<u>Proof</u>: As k is continuous, $V(k;I)$ is continuous in I and

(16.13) $V(k;E;X \cap I) = V(k;I;X) \quad (I \subseteq E)$.

Thus by (16.13) we can divide up E into a finite number of bricks for which $V(k;E;X \cap I) < \delta$, giving $V(h;E;X \cap I) < \varepsilon$, for suitable $\varepsilon > 0$, $\delta > 0$, as h is k-AC*. Hence h is VB* over X.

For (16.11) we use Theorem 16.4. We cannot replace AC* by ACG* in (16.11) as h has to be VB*.

For (16.12) with h integrable to H, and X the set of k-variation 0 where the derivative does not exist,

$$V(H-fk;E) \leq V(H;E;X) + V(fk;E;X) + V(H-fk;E;\sim X).$$

On the right the second V is 0 by Theorem 5.7 (5.22), in the first V,H is k-AC* in X and so that V is 0, and the third V is 0 by the definition of the derivative. Hence the result.

Let $k(I,x) \geq 0$, defined for E^c, be continuous there are integrable to K in the elementary set E. A function f being the limit k-almost everywhere of a sequence $(s_p(x))$ of step functions in E^c, is said to be k-measurable in E. Thus the conclusions of Theorems 16.1, 16.3 are that f is μ-measurable in E and k-measurable in E, respectively. The conclusions of Theorems 16.2, 16.4 are that the derivative $d(H;k;E;x)$ is k-measurable, where $k = \mu$ in Theorem 16.2.

We now look at another way of defining k-measurability. For a real point function f and constants $a < b$ let

$$X(f = a),\ X(f > a),\ X(f \geq a),\ X(f < a),\ X(f \leq a), X(a \leq f < b),\ \ldots$$

denote the sets of points x for which the appropriate equality or inequalities hold.

Theorem 16.6: *Let* $k(I,x) \geq 0$ *be defined on the closure* E^c *of the elementary set* E, *with* f,a,b *as above, such that* k, fk *and* $|f|k$ *are all integrable over* E. *Let* X *be any one of the sets mentioned above. Then* $k.\chi(X;.)$ *is integrable to* $V(k;E;X)$ *over* E.

Proof: For $X(f \geq b)$ see *Ex.* 8.4. Then $\chi(X(f > b);.)$ is the monotone increasing limit of $\chi(X(f \geq b+j^{-1})$ as $j \to \infty$, while

$$\chi(X(f < a);.) = \chi(X(-f > -a);.),\ \chi(X(f \leq a);.) = \chi(X(-f \geq -a);.),$$

$$\chi(X(a \leq f < b);.) = \min\,(\chi(X(f \geq a);.), \chi(X(f < b);.))$$

$$\chi(X(f = a);.) = \min\,(\chi(X(f \geq a);.), \chi(X(f \leq a)).$$

The integrability follows from Theorems 8.1 and 6.3. Similarly, omitting $\chi(X(f = a);.)$, from the integrability of any one of the other $\chi(X;.)k$ for an everywhere dense set of a,b, with $a < b$, we find the integrability of the others. This gives the second definition of f being k-measurable. Also we can say that a set X *is* k-*measurable* if $\chi(X;.)k$ is integrable.

To tie the two definitions of measurability together, we have the next theorem.

Theorem 16.7: *Let* $k(I,x) \geq 0$ *be defined for* E^c *and let* $f(x) \geq 0$ *in* E^c *be such that* $\chi(X(a \leq f < b);.)k$ *is integrable in* E *for all constants* a,b *satisfying* $0 \leq a < b$, *or all such constants in a set* Y *everywhere dense in* $[0,\infty)$. *Let* $(a_j) \subseteq Y$ *be a strictly increasing sequence with* $a_0 = 0 \in Y$ *and* $a_j \to \infty$ *as* $j \to \infty$, *and take* $g = a_j$ *in* $X(a_j \leq f < a_{j+1})(j = 0,1,2,\ldots)$. *Then* $g \leq f$, *and as the* χk *are integrable,* g *is integrable to*

$$(16.14)\qquad \sum_{j=1}^{\infty} a_j V(k;E;X(a_j \le f < a_{j+1}))$$

provided that this series is convergent. If divergent for a particular partition using the a_j*, the series is divergent for all partitions of this type, and* g *is not integrable. If we take a sequence of such partitions of* $[0,\infty)$ *(for once, having an infinity of partition points) in which for each* j*,* $a_{j+1}-a_j \to 0$*, then* $g \to f$*. Making each partition a refinement of the previous partition, the corresponding* g *are monotone increasing. If the integrals of the* g *all exist and the set of them is bounded with upper bound* G *then* f *itself is integrable with integral* G.

Proof: See Theorem 8.2.

Note that if in (16.14) we omit the terms after $j = N-1$, then $a_N = A$ (fixed) and if $a_{j+1}-a_j \to 0$ (all j) we have the original definition of Lebesgue for the integral of functions bounded by 0 and A, except that Lebesgue uses measure instead of variation. It is just as easy to use the limit of the infinite series (16.14) for the integral of non-negative functions f.

There is another definition of the measurability of sets. As we are keeping k and E fixed and varying X alone, for simplicity we can write V(X) for V(k;E;X). Then by Theorem 5.6

$$(16.15)\qquad V(Y) \le V(X \cap Y) + V(Y \setminus X),$$

for all sets X,Y in R^n. If for fixed X and every Y,

$$(16.16)\qquad V(Y) = V(X \cap Y) + V(Y \setminus X),$$

then we say that X *is a (Carathéodory)* k*-measurable set.*

Similarly we can look at the variation set of Section 7, writing VS(X) for the set $VS^c(k;E;X)$ which lies in $VS^c(k;E)$. If for some positive function δ on E^c, $VS(k;\delta;E)^c$ is compact, then by Theorem 7.6,

$$(16.17)\qquad VS(Y) \subseteq VS(X \cap Y) + VS(Y \setminus X),$$

where the right side is the collection of x+y for all $x \in VS(X \cap Y)$, all $y \in VS(Y \setminus X)$. If

$$(16.18) \quad VS(Y) = VS(X \cap Y) + VS(Y \setminus X),$$

for fixed X and all Y, then X is called *(Carathéodory)* k-*measurable* (using variation sets). Note the similarity between (16.15) and (16.17), between (16.16) and (16.18). This similarity carries through to proofs of some results of both series.

For the variation sets we assume (16.19), it is a simpler assumption than one having a different compact set for each X involved, though the second is more general.

(16.19) For some positive function δ on E^c let $VS(k;\delta;E)^c$ be compact.

Theorem 16.8: *If* X *is the union of a finite or infinite sequence* (X_j) *of mutually disjoint* k-*measurable sets, then* X *is* k-*measurable and for each set* $Y \subseteq R^n$,

$$(16.20) \quad V(Y) = \bigcup_{j=1}^{\infty} V(Y \cap X_j) + V(Y \setminus X).$$

If (16.19) *holds, we can put* VS *wherever* V *occurs in* (16.20).

Proof: We use induction. For all $Y \subseteq R^n$ let

$$(16.21) \quad V(Y) = \sum_{j=1}^{p} V(Y \cap X_j) + V(Y \setminus Z_p), \; Z_p \equiv \bigcup_{j=1}^{p} X_j, \; Z_p$$

k-measurable. Replacing Y by $Y \cap Z_p$ we get

$$(16.22) \quad V(Y \cap Z_p) = \sum_{j=1}^{p} V(Y \cap X_j)$$

so that in particular Z_p is k-measurable $(p = 1,\dots,q)$. Since X_{q+1} is k-measurable and disjoint from the k-measurable Z_q,

$$V(Y) = V(Y \cap X_{q+1}) + V(Y \setminus X_{q+1}) = V(Y \cap X_{q+1}) + V(Y \cap Z_q \setminus X_{q+1})$$

$$+ V(Y \setminus (X_{q+1} \cup Z_q)) = V(Y \cap X_{q+1}) + V(Y \cap Z_q) + V(Y \setminus Z_{q+1}).$$

From (16.22) for $p = q$ we have for $p = q+1$ the equality in (16.21) and so (16.22), and then Z_{q+1} is k-measurable. Thus (16.21) is true for each finite sequence, and true for all integers q when the original sequence is infinite, which we now assume.

Up to this point we can replace V by VS, the proof is the same. Then

$$V(Y \setminus X) \leq V(Y \setminus Z_p), \sum_{j=1}^{p} V(Y \cap X_j) + V(Y \setminus X) \leq V(Y) \leq \sum_{j=1}^{\infty} V(Y \cap X_j)$$

$$+ V(Y \setminus X).$$

Letting $p \to \infty$ we have (16.20), and by Theorem 5.6 X is k-measurable since

$$Y(Y \cap X) + V(Y \setminus X) \leq \sum_{j=1}^{\infty} V(Y \cap X_j) + V(Y \setminus X) = V(Y).$$

To finish the proof for VS,

$$VS(Y \setminus X) \subseteq VS(Y \setminus Z_p), \sum_{j=1}^{p} VS(Y \cap X_j) + VS(Y \setminus X) \subseteq VS(Y),$$

$$\sum_{j=1}^{<\infty} VS(Y \cap X_j) + VS(Y \setminus X) \subseteq VS(Y).$$

As VS(Y) is closed it contains the closure of the left side, which is the right side of (16.20) as $VS(Y \setminus X)$ is closed, using (16.19). Hence the theorem for VS.

If $Y \subseteq X$ with Y k-measurable and X arbitrary, then from the Carathéodory definition,

$$V(Y) + V(X \setminus Y) = V(Y \cap X) + V(X \setminus Y) = V(X),$$

(16.22) $\quad V(X \setminus Y) = V(X) - V(Y).$

But there seems to be no sensible subtraction operation for VS, and

$$VS(X \setminus Y) + VS(Y) = VS(X) \quad (Y \subseteq X)$$

does not always give the value of $VS(X \setminus Y)$. However, we can prove k-measurability.

<u>Theorem 16.9</u>: *If* $Y \subseteq X$ *with* X, Y *k-measurable, then* $X \setminus Y$ *is k-measurable.*

<u>Proof</u>: For V we have to show that if

$$(16.23) \quad P = Z \cap (X \setminus Y),\ Q = Z \setminus (X \setminus Y) = Z \cap (\setminus X \cup Y),\ V(P)+V(Q)$$
$$= V(P \cup Q) = V(Z).$$

To this end we put

$$Q_1 = Q \cap Y,\ Q_2 = Q \setminus Y,\ Q_1 \subseteq Y,\ Q_2 \subseteq \setminus Y,\ V(P)+V(Q) = V(P)+V(Q_1)+V(Q_2).$$

Noting that

$$P \subseteq X,\ Q_2 \subseteq (\setminus X \cup Y) \cap \setminus Y = \setminus X \cap \setminus Y \subseteq \setminus X,\ V(P)+V(Q_2) = V(P \cup Q_2).$$

To obtain (16.23) we need only note that

$$P = Z \cap X \cap \setminus Y \subseteq \setminus Y,\ P \cup Q_2 \subseteq \setminus Y,\ Q_1 \subseteq Y,\ V(P \cup Q_2)+V(Q_1) = V(P \cup Q).$$

As the proof only uses addition the result for VS follows in exactly the same steps.

In the definition of k-measurability of a set X using V or VS, X and $\setminus X$ appear symmetrically, so that if X is k-measurable, so is $\setminus X$. This with Theorems 16.8, 16.9, shows that the family M of k-measurable set contains $\setminus X$ if $X \in M$; if $(X_j) \subseteq M$ is a mutually disjoint sequence with union X, then $X \in M$; if $Y \subseteq X$ with $X, Y \in M$, then $X \setminus Y \in M$. We say that such a family is *countably additive* if it satisfies the first two results omitting the $X\setminus Y$ result, and including that the empty set is in the family. Here it is trivial that the given M contains the empty set.

If the V are replaced by the values of a measure, (16.14) gives a Lebesgue-type definition of the integral as the mesh of the partition tends to 0. Then Theorem 3.4 shows that the integral produced is included in the generalized Riemann integrals, which links up the last definitions of k-measurability using V with the earlier ones. There is probably a much simpler proof of the link.

Finally we can give a result linked with Theorem 11.1, and a small point in the proof of Lemma 10.1.

Theorem 16.10: *The function* M *of Theorem* 11.1 *can be* k*-measurable. If* (11.1) *is true, if* $f_j(x) \geq 0$ *for all* j,x, *and if the compact set* S *is fixed, the* k*-measurability of* M *implies that* (11.2) *is true.*

Proof: As $f_j k$ is integrable, f_j is k-measurable. (In dimension $n > 1$ we have to take $k = \mu$.) Hence so are f and $f_j - f$. Hence we can take M k-measurable. Conversely, with the given conditions, let $p_j(x)$ be the indicator of the set of x where $M(x) \leq j$. As M is k-measurable, so is p_j, and as f_j is integrable over E and $0 \leq f_j p_j \leq f_j$, then $f_j p_j$ is integrable. Further, $f_j p_j \to f$ everywhere as $j \to \infty$. If m is a positive integer valued function then $m(x) < M(x)$ gives $p_{m(x)}(x) = 0$, and $m(x) \geq M(x)$ gives $p_{m(x)}(x) = 1$. As $f_j \geq 0$ and $S \subseteq [A,B]$ for finite A,B, then for all δ-fine divisions $\mathcal{D}$ of E, (11.4) gives

$$0 \leq (\mathcal{D}) \Sigma f_{m(x)}(x) p_{m(x)}(x) k(I,x) \leq B,$$

and $f_j p_j$ satisfies (11.4) with no restrictions on the positive integer valued m. Thus f is integrable using Theorem 9.1, last part, which proves Theorem 16.10.

Also, in Lemma 10.1 each f_j is m-measurable by Theorem 16.1, in the sense that f_j is the limit almost everywhere of a sequence of step functions, so that the limit function f is also m-measurable, and Y_p is m-measurable and satisfies Carathéodory's condition on Y_p and $\setminus Y_p$, giving what is required.

Ex. 16.1. If h is the function of *Ex.* 15.3 then the ratios $h(I)/\mu(I)$ and $\mu(I)/(\mu(I)+h(I))$ are bounded, but the corresponding derivatives do not exist anywhere, showing that the proof of Theorem 16.4 needs the integrability of h, $|h|$, k, or something similar.

Ex. 16.2. For $0 < x < 1$ let $f(x) = y(\log_e y)^{-3/2}$ $(y = x^{-1})$ with $f(0) = 0$. Show that for

$$0 < t < \tfrac{1}{2}, \int_{[t,\frac{1}{2})} f d\mu = 2(\log_e 2)^{-\frac{1}{2}} - 2(\log_e(t^{-1}))^{-\frac{1}{2}} \to 2(\log_e 2)^{-\frac{1}{2}}$$
$$\text{as } t \to 0+,$$

so that $f\mu$ is integrable over $[0,\frac{1}{2})$ to $2(\log_e 2)^{-\frac{1}{2}}$ by the Cauchy limit, Theorem 5.14. Show that $s_p(2^{-p-1}) = 2^p.2(p\log_e 2)^{-\frac{1}{2}}$, so that if g majorizes the sequence $(s_p(x))$, g is not integrable over $[0,\frac{1}{2})$.

Ex. 16.3. Let F be a closed set in R^n so that $\sim F$ is open, and let $G(\varepsilon)$ be the open union of spheres with radius $\varepsilon > 0$ and centre each point of F. Show that as $\varepsilon \to 0$, $\sim G(\varepsilon) \to \sim F$, by noting that round each point of $\sim F$ there is a sphere free from points of F. For E the usual elementary set and a set $Y \subseteq E^c$, prove that

$$V(k;E;Y \cap F) + V(k;E;Y \sim G(1/j)) \leq V(Y),$$

by considering (I,x) in $G(1/(3j))$ and in a union of spheres with radius $1/(3j)$ and centres the points of $\sim G(1/j)$. Go on to prove that F is k-measurable by showing that the second V tends to $V(k;E;Y \sim F)$ as $j \to \infty$. Thus prove that G_δ-sets are k-measurable. *The outer* k*-measure* $k^*(X)$ of an arbitrary set $X \subseteq E^c$ is the smallest $V(k;E;H)$ of all G_δ-sets $H \supseteq X$. Show that

$$(16.24) \quad V(k;E;X) \leq k^*(X).$$

CHAPTER 6

CARTESIAN PRODUCTS AND THE FUBINI AND TONELLI THEOREMS

17 Fubini-type Theorems

Let m,n be positive integers with sum N. Then the co-ordinates $(x_1,\ldots,x_N)$ of points in R^N can be separated into two collections, one of m co-ordinates and one of n co-ordinates. The first collection can be any m of the N, provided that we take the same co-ordinate of each point in R^N, and then we can rearrange the co-ordinates so that the chosen m come first, $(x_1,\ldots,x_m)$. The second collection then becomes $(x_{m+1},\ldots,x_N)$, n in number. Let $X \subseteq R^m$, $Y \subseteq R^n$. Then $Z = X \times Y$ denotes the set in R^N formed of all (x,y) with $x \in X$ and $y \in Y$, and set $X \times Y$ is called the *Cartesian product of* X *and* Y. Thus $R^m \times R^n = R^N$. We can define bricks I^m, I^n, I and elementary sets E^m, E^n, E in R^m, R^n, R^N, respectively, and can define the integration process in the three spaces in the way of this book.

If $E = E^m \times E^n$ we have two kinds of results, respectively associated with the names of Fubini and Tonelli. In the first kind we integrate over E and find what this implies for integrals over E^m and E^n separately. In the second kind, without looking at E we consider when can E^m and E^n be interchanged, leaving the value of the double integral unaltered. As usual we use the Pythagorean distance in R^p (p = m,n,N), the square root of the sum of squares of co-ordinates. For the first kind of result we find connections between divisions of E and divisions of E^m and E^n, the idea behind the construction occurring in lectures on Lebesgue integration by J.C. Burkill and based on a paper by W.H. Young; but the construction is vastly different from that in Lebesgue integration.

<u>Lemma 17.1</u>: *Given the above definitions let* δ *be a positive function on* $E^c = E^{mc} \times E^{nc}$. *To each point* $x \in E^{mc}$ *with* $\delta_{1x}(y) = 2^{-\frac{1}{2}}\delta(x,y)$, *let there correspond a* δ_{1x}*-fine division* $\mathcal{D}(x)$ *of* E^n. *Then there is a positive function* δ_2 *on* E^{mc} *such that if* (I^m,x) *is* δ_2*-fine and* $(I^n,y) \in \mathcal{D}(x)$, *then* $(I^m \times I^n,(x,y))$, *written* $(I^m,x) \times (I^n,y)$, *is* δ*-fine.*

Proof: Let $\mathcal{D}(x)$ consist of (I^{nj},y^j) $(1 \leq j \leq r)$ and define

$$\delta_2(x) = \min_{1\leq j\leq r} 2^{-\frac{1}{2}} \delta(x,y^j) > 0.$$

If (I^m,x) is δ_2-fine and if $(I^{nj},y^j) \in \mathcal{D}(x)$ then $I^m \times I^{nj}$ lies in the N-dimensional sphere with centre (x,y^j) and radius

$$\surd\{\delta_2(x)^2 + \tfrac{1}{2}\,\delta(x,y^j)^2\} \leq \surd\{\tfrac{1}{2}\delta(x,y^j)^2 + \tfrac{1}{2}\delta(x,y^j)^2\} = \delta(x,y^j).$$

Thus $(I^m,x) \times (I^{nj},y^j)$ is δ-fine.

Theorem 17.2: *(Fubini-type) Let the real or complex valued* $k(I^m,x)$, $q(x;I^n,y)$ *be defined in* E^p $(p = m,n$, *respectively). If for* $I = I^m \times I^n$, $z = (x,y)$,

$$(17.1) \qquad h(I,z) \equiv h(I^m \times I^n,(x,y)) = k(I^m,x)q(x;I^n,y)$$

is integrable to H *over the partial sets of* E, *then*

$$Q(x) \equiv \int_{E^n} dq(x;I^n,y)$$

exists except for the $x \in X^m$ *with* $V(k;E^m;X^m) = 0$. *If we put* $Q(x) = 0$ *in* X^m, *the integral of* $Q(x)k(I^m,x)$ *exists in* E^m *with value* $H(E)$, *so that*

$$(17.2) \qquad H(E) \equiv \int_E d(kq) = \int_{E^m} \{\int_{E^n} dq(x;I^n,y)\}\, dk(I^m,x).$$

Note that (17.1) is unsymmetrical. If it also holds when x,y are interchanged, then for all x,y,I^m,I^n, if $k(I^m,x) \neq 0$,

$$k(I^m,x)q(x;I^n,y) = k_1(I^n,y)q_1(y;I^m,x),\ q(x;I^n,y)$$
$$= k_1(I^n,y)q_1(y;I^m,x)/k(I^m,x).$$

In the last equation the left side does not depend on I^m. Hence

$$q_1(y;I^m,x)/k(I^m,x)$$

is a function, say $f(x,y)$, of x,y only, unless $k_1(I^n,y) = 0$. Hence we have

$$(17.3) \quad h(I,z) = f(x,y)k(I^m,x)k_1(I^n,y),$$

which is also true when $k = 0$ or $k_1 = 0$, or both. (17.3) is a more traditional symmetric form, but the unsymmetrical form has been used, see Section 20.

Proof of theorem: Given $\varepsilon > 0$, let the positive function δ on E^c be such that

$$(17.4) \quad (\mathcal{D})\,\Sigma\,|h(I,z) - H(I)| < \varepsilon$$

for all δ-fine divisions $\mathcal{D}$ of E, which is possible by Theorem 5.3. By Lemma 17.1 there are suitable $\mathcal{D}^n(x)$ dividing E^n, and a positive function δ_2 for E^{mc}, such that the conclusion of Lemma 17.1 holds. Let $\mathcal{D}^m$ be a δ_2-fine division of E^m. Then

$$(17.5) \quad (\mathcal{D}^m)\,\Sigma\,|k(I^m,x)\ (\mathcal{D}^n(x))\ \Sigma\,q(x;I^n,y) - H(I^m \times E^n)| < \varepsilon.$$

Let X^{mj} be the set of x where for all positive functions δ on E^c there are at least two $2^{-\frac{1}{2}}\delta(x,y)$-fine divisions $\mathcal{D}_1{}^n(x)$, $\mathcal{D}_2{}^n(x)$ of E^n for which

$$(17.6) \quad |(\mathcal{D}_1{}^n(x))\,\Sigma\,q - (\mathcal{D}_2{}^n(x))\,\Sigma\,q| > 2^{-j}.$$

Using $\mathcal{D}_j{}^n(x)$ Lemma 17.1 gives a positive function $\delta_2{}^j$ $(j = 1,2)$ and we can take $\delta_2 = \min(\delta_2{}^1, \delta_2{}^2)$. Thus from (17.5), (17.6),

$$(\mathcal{D}^m)\,\Sigma|k(I^m,x)|\chi(X^{mj};x)2^{-j} < 2\varepsilon, \quad V(k;\delta_2;E^m;X^{mj}) \le 2^{j+1}\varepsilon.$$

As $\varepsilon > 0$ is arbitrary, and by Theorem 5.6,

$$(17.7)\quad V(k;E^m;X^{mj}) = 0,\quad V(k;E^m;X^m) = 0,\quad X^m \equiv \bigcup_{j=1}^{\infty} X^{mj}.$$

If $x \in\!\!\smallsetminus X^m$ then for each integer j there is a $\delta_j(x,y) > 0$ such that, by (17.4),

$$(17.8)\quad (\mathcal{D}^j)\Sigma|h(I,z) - H(I)| < 4^{-j},$$

$$(17.9)\quad |(\mathcal{D}_1{}^n(x))\,\Sigma\, q - (\mathcal{D}_2{}^n(x))\,\Sigma\, q| \le 2^{-j}\ (x \in\!\!\smallsetminus X^m)$$

for all δ_j-fine $\mathcal{D}^j$ and all $2^{-\frac{1}{2}}\,\delta_j(x,y)$-fine pairs of divisions $\mathcal{D}_1{}^n(x)$, $\mathcal{D}_2{}^n(x)$ of E^n, so that $(\mathcal{D}^n(x))\ \Sigma q$ is fundamental (E^n). By Theorem 4.3, q is integrable over E^n, say to $Q(x)$. In (17.5) we now have to replace $(\mathcal{D}^n(x))\ \Sigma q$ by its limit $Q(x)$ to finish the proof. Replacing δ_j by $\min(\delta_1,\delta_2,\ldots,\delta_j) > 0$, we can assume that δ_j is monotone decreasing in j. For $x \in\!\!\smallsetminus X^m$ let $\mathcal{D}^{nj}(x)$ be a $2^{-\frac{1}{2}}\,\delta_j$-fine division of E^n with

$$f_j(x) \equiv (\mathcal{D}^{nj}(x))\ \Sigma\ q(x;I^n,y).$$

By (17.9) with δ_j monotone decreasing in j so that $\mathcal{D}^{nr}(x)$ is $2^{-\frac{1}{2}}\,\delta_j$-fine $(r \ge j)$,

$$(17.10)\quad |f_j(x) - f_r(x)| \le 2^{-j}\ (r > j),\ |f_j(x) - Q(x)| \le 2^{-j}\ (x \in\!\!\smallsetminus X^m).$$

Replacing $\mathcal{D}^n(x)$ in (17.5) with $\varepsilon = 4^{-j}$, by $\mathcal{D}^{nj}(x)$ and by $\mathcal{D}^{n,(j+1)}(x)$, with δ_2 to fit in with these divisions, and $\mathcal{D}^m$ any δ_2-fine division of E^m,

$$(17.11)\quad (\mathcal{D}^m)\ \Sigma\ |k(I^m,x)|\ |f_{j+1}(x) - f_j(x)| < 2.4^{-j}.$$

From (17.11), with X^{mjp} the set in $\smallsetminus X^m$ where

$$|f_{j+1}(x) - f_j(x)| > 2^{-j-p},$$

$$2^{-j-p}V(k;E^m;X^{mjp}) \le V(k|f_{j+1}-f_j|;E^m;X^{mjp}) \le 2.4^{-j},$$

$$(17.12)\quad V(k;E^m;X^{mjp}) \le 2^{1-j+p},\ V(k;E^m;Y^{mp}) \le 2^{1+p}\ (Y^{mp} = \bigcup_{j=1}^{\infty} X^{mjp}).$$

Using (17.7) and putting

$$s(x) = 2^{-1-2p}(x \in Y^{mp} \setminus (Y^{m1} \cup Y^{m2} \cup \ldots \cup Y^{m,(p-1)})),$$

$$(17.13)\qquad V(sk;E^m;Z) \leq 1, \quad Z = \bigcup_{p=1}^{\infty} Y^{mp} \cup X^m.$$

If $x \mathrel{\overline{\in}} Z$ then $x \mathrel{\overline{\in}} X^{mjp}$ for any j,p, and so

$$(17.14)\qquad |f_{j+1}(x) - f_j(x)| \leq 2^{-j-p}(\text{all } j,p), \; |f_{j+1}(x) - f_j(x)| = 0,$$

$$f_j(x) = f_1(x) = Q(x)$$

for $x \mathrel{\overline{\in}} Z$ and all j. By (17.10), given $\varepsilon > 0$, in $Z \setminus X^m$ we can choose the integer $t = t(x)$ so large that

$$(17.15)\qquad |f_{t(x)}(x) - Q(x)| < s(x)\varepsilon,$$

while $\mathcal{D}^{nt}(x)$ is $2^{-\frac{1}{2}}\delta_j(x,y)$-fine. By Lemma 17.1 we can find suitable δ_2, $\mathcal{D}^m$ to fit with $\mathcal{D}^{nj}(x)$ ($x \mathrel{\overline{\in}} Z$ and $x \in X^m$), $\mathcal{D}^{nt(x)}(x)$ ($x \in Z \setminus X^m$), then by (17.13), (17.14), (17.15) the replacement can take place.

The form (17.1) seems to be the most general for which the conclusion of Theorem 17.2 can be proved. But, assuming a little more, we can deal with general h.

<u>Theorem 17.3</u>: *Let* $I = I^m \times I^n$ *be the bricks in* R^N, *and* $E = E^m \times E^n$, *as before. For* $z = (x,y)$ *and* $h(I,z)$ *defined for* E^c, *let* h *be integrable over* E *to* H, *and for each fixed* (I^m,x) *let* $h(I^m \times I^n,(x,y))$ *be integrable relative to* (I^n,y) *over* E^n *to* $G(I^m,x)$. *Given* $\varepsilon > 0$ *let there be a* $k(I^m,x)$ *defined for* E^{mc} *and of variation less than* ε *over* E^m, *and a positive function* δ^n *on* E^{nc}, *such that*

$$(17.16)\qquad |G(I^m,x) - (\mathcal{D}^n)\textstyle\sum h(I^m \times I^n,(x,y))| \leq |k(I^m,x)|$$

for all δ^n-fine divisions $\mathcal{D}^n$ of E^n. Then there exists

$$(17.17)\qquad \int_{E^m} dG(I^m,x) \equiv \int_{X^m} d\{\int_{E^n} dh(I^m \times I^n,(x,y))\} = \int_E dh = H.$$

Proof: There is a positive function δ^m on E^{mc} such that

$$(17.18)\qquad (\mathcal{D}^m)\ \Sigma|k(I^m,x)| < V(k;E^m) + \varepsilon < 2\varepsilon,$$

for all δ^m-fine divisions $\mathcal{D}^m$ of E^m. There is a positive function δ on E^c such that (17.4) is true. For each $x \in E^{mc}$ let

$$\delta_1{}^n(y;x) = \min(2^{-\frac{1}{2}}\delta(x,y),\ \delta^n(y)) > 0$$

and let $\mathcal{D}_1{}^n(x)$ be a $\delta_1{}^n$-fine division of E^n. Let $\delta_2{}^m$ be a positive function on E^{mc} that corresponds to the various $\mathcal{D}_1{}^n(x)$ and the δ_n as in Lemma 17.1 and let

$$\delta_1{}^m = \min(\delta_2{}^m,\ \delta^m) > 0.$$

Then a $\delta_1{}^m$-fine division $\mathcal{D}_1{}^m$ of E^m satisfies

$$|(\mathcal{D}_1{}^m)\ \Sigma G - H| \leq |(\mathcal{D}_1{}^m)\ \Sigma G - (\mathcal{D}_1{}^m)\ \Sigma(\mathcal{D}_1{}^n)\ \Sigma h| + |(\mathcal{D}_1{}^m)\Sigma(\mathcal{D}_1{}^n)\ \Sigma h - H|$$
$$\leq (\mathcal{D}_1{}^m)\ \Sigma|k| + \varepsilon$$

by (17.16) and (17.4), since the division of E consisting of the Cartesian products of the $(I^m,x) \in \mathcal{D}_1{}^m$ with the $(I^n,y) \in \mathcal{D}_1{}^n$, is δ-fine by Lemma 17.1, and then by (17.18),

$$|(\mathcal{D}_1{}^m)\ \Sigma G - H| < 3\varepsilon,$$

proving (17.17).

If, for a given positive function δ^n, we define $|k(I^m,x)|$ as being the left side of (17.16) with $\mathcal{D}^n$ a δ^n-fine division of E^n, and if (17.17) is true, then the variation of k is arbitrarily small by (17.4).

Ex. 17.1: Let f(x) be a point function on the real line and let H be a finitely additive function of rectangles $I \times J$ for intervals I and J on the real line. Show that if fH is integrable on $T = [a,b) \times [u,v)$,

$$\iint_T f\,dH = \int_{[a,b)} f(x)\,dH(I \times [u,v)).$$

Proof: Let δ be a positive function on T^c such that for all δ-fine divisions $\mathcal{D}$ over T,

$$\left|(\mathcal{D}) \Sigma f(x)H(I \times J) - \iint_T f\,dH\right| < \varepsilon.$$

Let $\mathcal{D}^n(x)$ be a $2^{-\frac{1}{2}}\delta(x,y)$-fine division of $[u,v)$ for each $x \in [a,b]$, with $\delta_2 > 0$ on $[a,b]$ as in Lemma 17.1, and $\mathcal{D}^m$ a δ_2-fine division of $[a,b)$ $(m = 1 = n)$. Then

$$\left|(\mathcal{D}^m) \Sigma f(x)(\mathcal{D}^n(x)) \Sigma H(I \times J) - \iint_T f\,dH\right| < \varepsilon,\ (\mathcal{D}^n(x))\Sigma H(I \times J)$$
$$= H(I \times [u,v)).$$

Ex. 17.2: Let $h(x,J)$ be finitely additive in $J \subseteq [u,v)$, with $k = k(I)$, $I \subseteq [a,b)$, where the brick-point pair in R^2 is $(I,x) \times (J,y)$. If hk is integrable over $[a,b) \times [u,v)$, show that the integral is equal to

$$\int_{[a,b)} h(x;[u,v))\,dk(I).$$

Ex. 17.3: Let h,k be VB* in intervals $[a,b]$, $[u,v]$, respectively, on the real line, and let W be a set in R^2 with indicator χ, with W_x the set of y for the fixed x, such that $(x,y) \in W$, the section of W for the constant x, and with W_y the section of W for the constant y. If $|hk|\chi$ is integrable over $[a,b) \times [u,v) = T$ then

$$V(hk;T;W) = \int_{[a,b)} V(k;[u,v);W_x)\,d|h| = \int_{[u,v)} V(h;[a,b);W_y)\,d|k|.$$

Hint: Put the variations as integrals.

Ex. 17.4: Let $f(y)$ and $h(x;J)$ be given in $[u,v]$ for all $x \in [a,b]$, with

$k(I)$ VBG* in $[a,b]$. If fhk is integrable on T to $H(T)$ then

$$J(x) \equiv \int_{[u,v)} f\,dh(x;J)$$

exists except for x in a set X of k-variation zero. If we write 0 for $J(x)$ in X then Jk is integrable over $[a,b)$ to $H(T)$. If also $|fhk|$ is integrable on T then

$$J^+(x) \equiv \int_{[u,v)} |f|\,dV(h;.)$$

exists except for x in a set X of k-variation zero, and, writing 0 for J^+ in X, $J^+V(k;.)$ is integrable to $V(fhk;T)$ in $[a,b)$.

18 Tonelli-type Theorems And The Necessary And Sufficient Conditions For Reversal Of Order Of Double Integrals

Here we have the second kind of problem, the section's title being a good description. In symbols, taking integration over E^m to be the integration of Theorems 13.7, 13.8, and integration over E^n to be the limit process $\lim_Y$, we look at

$$(18.1)\qquad \int_{E^m} \{\int_{E^n} f(x,y)dk\}dh \equiv \{\int_{E^m} (\lim_{\delta\to 0+} (\mathcal{D}^n)\, \Sigma\, f(x,y)k(I^n,y))dh =$$

$$\lim_{\delta\to 0+} (\mathcal{D}^n)\, \Sigma\, k(I^n,y) \int_{E^m} f(x,y)dh \equiv \int_{E^n} \{\int_{E^m} f(x,y)dh\}dk,$$

where $\mathcal{D}^n$ is a general δ-fine division of E^n and where $\lim_{\delta\to 0+}$ is the limit as δ shrinks.

Here, monotonicity does not seem relevant, but the analogue of the bounded Riemann sums test is very relevant and leads to the following theorem.

<u>Theorem 18.1</u>: (18.2) *For* E^m *an elementary set,* $h(I,x) \geq 0$ *on* E^{mc}, *and* $f(x,y)$ *real-valued for* $x \in E^{mc}$, $y \in E^{nc}$, *let* $f(x,y)h(I^m,x)$ *be integrable*

over E^m *for each fixed* $y \in E^{nc}$, *so that if* $\mathcal{D}^n$ *is a fixed division of* E^n, *the* $f(x,y)$ *of Theorems* 13.7, 13.8 *is replaced by*

$$(18.3)\quad (\mathcal{D}^n)\ \Sigma f(x,y)k(I^n,y)$$

which when multiplied by $h(I^m,x)$ *is integrable over* E^m. *We also assume that* (18.3) *is integrable over* E^n *for each fixed* $x \in E^{mc}$.

(18.4) *Let* $\mathcal{D}^m$ *be a positive function on* E^{mc}, *let* $B < C$ *be two real numbers, and let*

$$B \leq (\mathcal{D}^m)\ \Sigma\{(\mathcal{D}^n(x))\,\Sigma f(x,y)k(I^n,y)\}h(I^m,x) \leq C$$

for all δ^m*-fine divisions* $\mathcal{D}^m$ *of* E^m *and all collections of divisions* $\mathcal{D}^n(x)$ *of* E^n. *Then* (18.1) *holds.*

Proof: The step function approximation of the integral over E^n, from Theorems 16.1, 16.3, needs the limitation that k is continuous or even the volume function, until these theorems can be improved. However, for a step function $s_p(y)$, if $\mathcal{D}^n(x)$ is a refinement of the partition using the brick "steps", the sum over $\mathcal{D}^n(x)$ is equal to

$$\int_{E^n} f(x,y)dk,$$

and (18.4) simplifies to

$$B \leq (\mathcal{D}^m)\ \Sigma h(I^m,x) \int_{E^n} f(x,y)dh \leq C$$

which is true for (18.4) as it stands. We now use Theorem 9.1 to finish the proof.

We now come to the necessary and sufficient conditions for (18.1).

Theorem 18.2: *Let* $h(I^m,x)$ *and* $k(I^n,y)$ *be defined and* VBG*, *respectively for* E^{mc}, E^{nc}, *with* (18.2). *In order that*

(18.5) $\int_{E^m} \{\int_{E^n} f dk\} dh = F$

exists it is necessary and sufficient that for some compact set S *of arbitrarily small diameter, some positive function* δ^m *on* E^{mc} *, some positive function* $\delta^n(y;x)$ *on* E^{nc} *for each* x *in* E^{mc}*, all* δ^m*-fine divisions* $\mathcal{D}^m$ *of* E^m *and all* $\delta^n(y;x)$*-fine divisions* $\mathcal{D}^n(x)$ *of* E^n,

(18.6) $(\mathcal{D}^m)\ \Sigma h(I^m,x)(\mathcal{D}^n(x))\ \Sigma f(x,y)k(I^n,y) \in S.$

Given (18.6), *a necessary and sufficient condition that* (18.1) *holds is that for each* $\varepsilon > 0$ *there is a positive function* δ^n *on* E^{nc} *such that for all* δ^n*-fine divisions* $\mathcal{D}^n$ *of* E^n*, a positive function* $\delta^m(x;y)$ *on* E^{mc} *and depending on* δ^n*, all* $\delta^m(x;y)$*-fine divisions* $\mathcal{D}^m(y)$ *of* E^m*, and a compact set* T *of diameter* $< \varepsilon$ *such that* $S \cap T$ *is not empty, with*

(18.7) $(\mathcal{D}^n)\ \Sigma\{(\mathcal{D}^m(y))\ \Sigma f(x,y)h(I^m,x)\}k(I^n,y) \in T.$

<u>Proof</u>: If (18.5) exists then, given $\varepsilon > 0$, there is a positive function δ^m on E^{mc} with

(18.8) $F-\varepsilon < (\mathcal{D}^m)\ \Sigma h(I^m,x) \int_{E^n} f dk < F + \varepsilon.$

As h is VBG* there is a positive function s(x) such that sh is VB* (Theorem 5.9) so that we can also assume that for some constant $M > 0$,

(18.9) $(\mathcal{D}^m)\ \Sigma s\ |h| \le M.$

Also there is a positive function $\delta^n(y,x)$ on E^{nc} such that for each δ^n-fine division $\mathcal{D}^n(x)$ of E^n,

(18.10) $|(\mathcal{D}^n(x))\ \Sigma f(x,y)k(I^n,y) - \int_{E^n} f dk| < \varepsilon s(x).$

Putting together (18.8), (18.9), (18.10) we have (18.6) with $S = [F - \varepsilon - \varepsilon M, F + \varepsilon + \varepsilon M]$. Conversely, if (18.6) is true, then as $\delta^n(y,x) \to 0$ we have

$$(\mathcal{D}^m)\ \Sigma h(I^m,x) \int_{E^n} f dk \in S$$

and (18.5) exists as S has arbitrarily small diameter. Next, if

$$(\mathcal{D}^n)\ \Sigma k(I^n,y) \int_{E^m} f(x,y) dh = \int_{E^m} \{(\mathcal{D}^n)\ \Sigma f(x,y)k(I^n,y)\} dh$$

tends to F as $\delta^n \to 0$ on E^{nc} with δ^n-fine $\mathcal{D}^n$ then, given $\varepsilon > 0$,

$$F-\varepsilon < \int_{E^m} \{(\mathcal{D}^n)\ \ \Sigma f(x,y)k(I^n,y)\}\ dh < F + \varepsilon,$$

and using the analogue of (18.10), for a suitable positive function $\delta^m(x;y)$ on E^m and all $\delta^m(x;y)$-fine divisions $\mathcal{D}^m(y)$ of E^m,

$$F - \varepsilon < (\mathcal{D}^n)\ \Sigma k(I^n,y)(\mathcal{D}^m(y))\ \Sigma f(x,y)h(I^m,x) < F + \varepsilon.$$

Thus we can take $T = [F - \varepsilon,\ F + \varepsilon]$, of diameter 2ε, while S and T have the point F in common.

Conversely, let $\delta^m(x;y) \to 0+$ such that the inner sum tends to $\int_{E^m} f dh$. Then

$$(\mathcal{D}^n)\ \Sigma k(I^n,y) \int_{E^m} f dh \in T.$$

As T has an arbitrarily small diameter, the last sum tends to a limit

$$F^* = \int_{E^n} \left\{\int_{E^m} f dh\right\} dk\ .$$

As S,T always have points in common, $F^* = F$ and we have proved the result.

Let us now take integration over E^n to be the integration of Theorems 13.7, 13.8, and integration over E^m to be the limit process $\lim_{Y}$, i.e. let us interchange the roles of the two integration processes. Then it is easy to see that the conditions (18.6) and (18.7) interchange, which gives a check of the theory.

We have replaced f(x,y) in the general theory of Section 13 by (18.3). If we do the reverse operation in Section 17, replacing expressions like (18.3) by f(x,y) or by $f_j(x)$ we obtain an analogue of the double integral, involving sequences of integrals.

For $j = 1,2,\ldots$ let $f_j k$ be integrable over the elementary set E. If there is a number F such that, given $\varepsilon > 0$, there are an integer J and positive functions δ_j on E^c with

$$(18.11) \quad F - \varepsilon < (\mathcal{D}) \Sigma f_{j(x)}(x)h(I,x) < F + \varepsilon$$

whenever the integer-valued function $j(x) \geq J$ and $\mathcal{D}$ is a $\delta_{j(x)}$-fine division of E, then we say that $f_{j(x)}(x)h(I,x)$ is integrable to F over E.

Theorem 18.3: *If* h *is* VBG* *in* E^c *and for each fixed integer* j, $f_j h$ *is integrable in* E *and if* $f_{j(x)}(x)h$ *is integrable over* E, *then for some set* X *of variation zero and for* χ *the indicator of* $\backslash X$, $f_j\chi$ *satisfies* (11.1), (11.2), (11.3), *except that* h *is* VBG*, *not* VB*.

Proof: Writing F as F(E), by the usual proofs in Section 5 the integral $F(E_1)$ exists over every partial set E_1 of E, and is finitely additive, and for every $j(x) \geq J$ and every $\delta_j(x)$-fine division $\mathcal{D}$ of E,

$$(18.12) \quad (\mathcal{D}) \Sigma |f_{j(x)}(x)h(I,x) - F(I)| \leq 8\varepsilon.$$

If $q(x)$ is another integer-valued function on E^c with $q(x) \geq J$, if $\delta = \min (\delta_{j(x)}(x), \delta_{q(x)}(x))$ and if $\mathcal{D}$ is a δ-fine division of E, then

$$(18.13) \quad (\mathcal{D})\Sigma|f_{j(x)} - f_{q(x)}(x)| \, |h(I,x)| \leq 16\varepsilon.$$

Let $X_r \subseteq E^c$ be the set of all x in E^c having the property that, given J, there are two integers j,q in $q > j \geq J$ and depending on x, such that

$$(18.14) \quad |f_j(x) - f_q(x)| \geq 1/r \ (r = 1,2,\ldots)$$

writing $j = j(x)$, $q = q(x)$, if χ is the indicator of X_r, and X the union of the X_r, for every δ-fine division $\mathcal{D}$ of E, (18.13), (18.14) give

$$(\mathcal{D})\Sigma|h(I,x)|\chi \leq 16r\varepsilon, \quad V(h;E;X_r) \leq 16r\varepsilon.$$

This is true for every J, so that we can let $\varepsilon \to 0+$ and have

$$V(h;E;X_r) = 0 = V(h;E;X).$$

If $x \in \sim X$, then given the integer $r > 0$, there is an integer J_r such that

$$|f_j(x) - f_q(x)| < 1/r \quad (\text{all } q > j \geq J_r)$$

and $(f_j(x))$ is a fundamental and so convergent sequence in j, with limit f(x), say. Thus if χ^* is the indicator of $\sim X$ we can replace f_j by $f_j\chi^*$, leaving integrability unaltered, together with the integrals, and the new $f_j \to f$ everywhere, and fh is integrable on E to F. As h is VBG* there is a function $s(x) > 0$ such that $V(sh;E) \leq 1$. Given $\varepsilon > 0$, there is an integer-valued function $p \geq J$ on E^c such that

$$|f_{p(x)}(x) - f(x)| < \varepsilon s(x).$$

For $\mathcal{D}$ a $\delta_{p(x)}$-fine division of E, (18.11) gives

$$|(\mathcal{D})\,\Sigma\, f(x)h(I,x) - F\,| \leq \varepsilon + \varepsilon(\mathcal{D})\,\Sigma\,|sh| < 3\varepsilon\,.$$

Finally, taking j(x) constant at $j \geq J$ in (18.11) with $\mathcal{D}$ δ_j-fine, and then by the definition of the integral F_j of f_jh over E using a δ_j^*, we can replace δ_j by $\delta_j^{o} = \min(\delta_j, \delta_j^*)$ and let $\mathcal{D}$ be δ_j^{o}-fine, so that (18.11) is true and

$$|(\mathcal{D})\,\Sigma\, f_j(x)h(I,x) - F_j| < \varepsilon, \; |F_j - F| < 2\varepsilon \quad (j \geq J)$$

implying that $F_j \to F$ as $j \to \infty$. This completes the proof.

Note that (18.11) and (11.4) are effectively the same. However, for (18.11) we assume $j(x) \geq J$, where J is a constant, and for (11.4) we assume $j(x) \geq M(x)$, and M(x) need not be constant.

Ex. 18.1: Let $f(x,y) = 2(x-y)(x+y)^{-3}$ $(x \geq 0,\ y \geq 0,\ x+y > 0)$, $f(0,0) = 0$.

Then as the calculus integral is a generalized Riemann integral, keeping y fixed,

$$\int_{[0,1)} f(x,y)d\mu = [-2z^{-1} + 2yz^{-2}]_y^{1+y} \; (z = x+y), = -2(1+y)^{-2}.$$

Integrating this relative to y over [0,1) gives -1. Further, noting that $f(x,y) = -f(y,x)$, integrating over y first and then x, we get 1. Thus this example cannot satisfy the necessary and sufficient conditions of Theorem 18.2, and it cannot satisfy the conditions of Theorem 17.2, so that we cannot integrate it, multiplied by the volume function, over $[0,1) \times [0,1)$. Can these two results be proved directly? Let x and y lie in (0,1), and let $L(x,y)$ be the main square, $[0,1) \times [0,1)$ minus the rectangle $[0,x) \times [0,y)$, so that

$$L(x,y) = L(1,y) \cup \{[x,1) \times [0,y)\}.$$

Show that the integral over $L(x,y)$ is $(y-x)(y+x)^{-1}$. If $x \to 0$ first, and then $y \to 0$, the limit is 1, while if $y \to 0$ first, and then $x \to 0$, the limit is -1.

Ex. 18.2: In R^2 let $f(x,y) = x/y$ when $y > 0$, and $f(x,0) = 0$. Show that there exists

$$\int_{[0,1)} \{\int_{[-1,1)} f(x,y)dx\}dy, \text{ but that } \int_{[0,1)} \{\int_{[0,1)} f(x,y)dx\}dy$$

does not exist, so that the repeated integral does not have one property of the integral in R^2. Does this imply anything about the latter integral?

CHAPTER 7
APPLICATIONS

19 Ordinary Differential Equations

Given an (n+1)-dimensional open set G of points $(x,\underline{y})$, x being a real number and $\underline{y}$ an n-dimensional vector, let $(a,\underline{b}) \in G$. We study differential equations of the type

$$(19.1) \quad d\underline{y}/dx = \underline{g}(x,\underline{y}),$$

where $\underline{g}$ is a function whose values are finite n-dimensional vectors. We study the existence of a solution $\underline{y} = \underline{f}(x)$ with $\underline{f}(a) = \underline{b}$, such that $(x,\underline{y}) \in G$ when $|x-a| \leq \alpha$, for some $\alpha > 0$. We can assume that $x \geq a$ as the same methods can be used when $x \leq a$. The (generalized Riemann) integral integrates all finite-valued derivatives, so that we can integrate (19.1) to obtain

$$(19.2) \quad \underline{y} - \underline{b} = \int_{[a,x)} \underline{g}(x,\underline{y})dx,$$

where by dx we mean $d\mu$ with $\mu([u,v),x) = v-u$. Conversely we obtain (19.1) (at least, almost everywhere) on differentiating (19.2).

In (19.2) we can relax the conditions on $\underline{g}$ slightly, and $\underline{g}(x,\underline{f}(x))$ can behave badly in a set of x of variation zero without altering (19.2), whatever conditions are later imposed on $\underline{g}$.

Beginning trivially, if $\underline{y} = \underline{b}$ is a solution through $(a,\underline{b})$, then by (19.2) the integral of $\underline{g}(x,\underline{b})$ is $\underline{0}$ in some neighbourhood of a, so that $\underline{g}(x,\underline{b})$ itself is $\underline{0}$ almost everywhere in that neighbourhood. Conversely, if $\underline{g}(x,\underline{b}) = \underline{0}$ almost everywhere then $\underline{y} = \underline{b}$ is a solution through $(a,\underline{b})$. There may be other solutions through $(a,\underline{b})$.

Carathéodory showed that at times (19.2) has a solution, and here we go far beyond the cases he treated, supposing that for some $\alpha > 0$ the method works over the range $[a,a+\alpha]$, writing

$$(19.3)\quad \underline{f}_j(x) = \begin{cases} \underline{b} & (a \le x \le a+\alpha/j), \\ \underline{b} + \int_{[a,x-\alpha/j)} \underline{g}(t,\underline{f}_j(t))dt & (a+\alpha/j \le x \le a+\alpha). \end{cases}$$

Here, $\underline{f}_j$ is defined as a continuous function in $(a+p\alpha/j, a+(p+1)\alpha/j]$ by using its values in $[a,a+p\alpha/j]$. This is an inductive definition for $p = 1,2,\ldots,j-1$. Of course $\underline{f}_j$ is not usually a solution of (19.2) because of $x-\alpha/j$ in the integral of (19.3). Carathéodory's idea is to take a subsequence of j tending to infinity, for which the $\underline{f}_j$ tend to a limit $\underline{y} = \underline{f}$, which would be a solution of (19.2) if the limit could be taken inside the integral and inside $\underline{g}$. To this end he assumes that $\underline{g}(x,\underline{y})$ is continuous in $\underline{y}$ for almost all x, and that for a suitable modulus or norm $|\cdot|$ in R^n, some integrable $m(x) \ge 0$, all $\underline{y}$, and almost all x,

$$(19.4)\qquad |\underline{g}(x,\underline{y})| \le m(x).$$

Then the $\underline{f}_j(x)$ are uniformly bounded and equicontinuous, and by Ascoli's theorem a subsequence is uniformly convergent on $[a,a+\alpha]$, and gives a solution.

A more general condition for limits under the integral sign is given in Theorem 9.1, last part. Here the condition is as follows. For a compact set S in R^n, some positive function δ on $[a,a+\alpha]$, all δ-fine divisions $\mathcal{D}$ of $[a,a+\alpha)$, and all $\underline{f} : [a,a+\alpha] \to R^n$,

$$(19.5)\qquad (\mathcal{D})\ \Sigma\ \underline{g}(x,\underline{f}(x))(v-u) \in S.$$

If S lies in an n-dimensional cube with edge length A, then as in the proof of Theorem 9.4,(19.5) gives

$$(19.6)\qquad (\mathcal{D})\ \Sigma|\underline{g}(x,\underline{f}(x)) - \underline{g}(x,\underline{b})|\ (v-u) \le 2nA.$$

In this we put the $\underline{f}_j$ of (19.3) for $\underline{f}$ and take δ^*-fine divisions on each $[u-\alpha/j, v-\alpha/j)$ with $([u,v),x) \in \mathcal{D}$ and let δ^* shrink suitably. Thus

$$(19.7)\quad (\mathcal{D})\,\Sigma\left|\underline{f}_j(v) - \underline{f}_j(u) - \int_{[u-\alpha/j,v-\alpha/j)} \underline{g}(t,\underline{b})dt\right|$$

$$= (\mathcal{D})\Sigma\left|\int_{[u-\alpha/j,v-\alpha/j)} \{\underline{g}(t,\underline{f}_j(t)) - \underline{g}(t,\underline{b})\}dt\right| \le 2nA,$$

provided that one division-point of $\mathcal{D}$ is $a+\alpha/j$. Below this number, terms of the sum are 0. Thus

$$(19.8)\quad \underline{f}_j(x) - \int_{[a,x-\alpha/j)} \underline{g}(t,\underline{b})dt$$

is of uniform bounded variation in x and j. As (19.8) is 0 at x = a, Theorem 2.10 Corollary shows that a subsequence of (19.8) tends to a finite limit. By continuity of the integral of $\underline{g}(t,\underline{b})$, the same subsequence of $(\underline{f}_j(x))$ tends to a finite limit, say $\underline{f}(x)$, everywhere in $[a,a+\alpha]$. Assuming that $\underline{g}(x,\underline{y})$ is continuous in $\underline{y}$ for almost all x, the same subsequence of $(\underline{g}(x,\underline{f}_j(x)))$ tends to $\underline{g}(x,\underline{f}(x))$ almost everywhere. Then (19.5) and Theorem 9.1 show that for the same subsequence, the limit can be taken under the integral sign, $\underline{y} = \underline{f}(x)$ is a solution of (19.2), and $\underline{f}$ is continuous.

We have thus proved the following result.

Theorem 19.1: *For a compact set* S *in* R^n, *some positive function* δ *on* $[a,a+\alpha]$, *all* δ*-fine divisions* $\mathcal{D}$ *of* $[a,a+\alpha)$, *and all* $f:[a,a+\alpha] \to R^n$, *let* (19.5) *be true, and let* $\underline{g}(x,\underline{y})$ *be continuous in* y *for almost all* x. *Then there is a continuous solution* $\underline{y} = \underline{f}(x)$ *of* (19.2) *through the point* (a,b) *in* R^{n+1}.

The result (9.5) in Theorem 9.1 is useful and is stronger than the Arzèla-Lebesgue (19.4), and is very near to the set of necessary and sufficient conditions of Section 11. So the question arises of whether (19.5) can be replaced by modifications of the necessary and sufficient conditions of Theorems 11.1, 11.2. We could replace the $\underline{f}_j(x)$ of these theorems by $\underline{g}(x,\underline{f}_{r_j}(\underline{x}))$ with $\underline{f}_j(x)$ given by (19.3) and with (r_j) an arbitrary subsequence of the positive integers, chosen so that $(\underline{f}_{r_j})$ is convergent everywhere, if this is possible. These seem to be the widest conditions for which a Carathéodory-type proof works using (19.3). However, it seems that the resulting modifications of (11.4), (11.9) would be too complicated and impractical, though they could easily be written if ever needed.

There are simple solvable differential equations that are not covered by the above proofs, for example, $\underline{g}(x,\underline{y}) = \underline{g}(x)/h(\underline{y})$ where $h : R^n \to R$ and $h(\underline{y}) \to \infty$ as $\underline{y}$ tends to some vector $\underline{b}$, where $\underline{g}(x)$ is integrable in some neighbourhood of $x = a$, and where $h(\underline{y})$ is integrable in some neighbourhood of $\underline{y} = \underline{b}$. Then (19.1) becomes

$$h(\underline{y})d\underline{y}/dx = \underline{g}(x), \int h(\underline{y})d\underline{y} = \int \underline{g}(x)dx$$

in the respective neighbourhoods of $\underline{b}$ and a, by integration by substitution (Theorem 5.7 (5.25)) on each co-ordinate of $\underline{y}$ separately. For example,

$$h(y_1,\ldots,y_n) = (y_1^p,\ldots,y_n^p) \text{ (p constant in } -1 < p < 0), \ \underline{b} = (0,\ldots,0).$$

To combine the two kinds of results let there be an integrable $h(\underline{y})$ for which $\underline{g}(x,\underline{y})h(\underline{y})$ satisfies our generalization of Carathéodory's conditions. Then

$$\int h(\underline{y})d\underline{y} = \int h(\underline{y})(d\underline{y}/dx)dx = \int h(\underline{y})\underline{g}(x,\underline{y})dx$$

and there is a solution.

The solution is unique if

(19.8) $\quad |\underline{g}(x,\underline{y}_1) - \underline{g}(x,\underline{y}_2)| \le L(|\underline{y}_1 - \underline{y}_2|)$

where $L(r)$ is continuous in $0 \le r \le k$ and some $k > 0$, $L(0) = 0, L(r) > 0$ except at 0, and $\int_{[0,k)} dr/L(r)$ does not exist. Also

(19.9) $L(r)$ can be replaced by $\phi(x)L(r)$ if also $\phi \ge 0$ is integrable on $[a,a+\infty)$.

Applied mathematics, or natural philosophy, is the application of pure mathematics to the natural world. The scientist's experiments or observations, and data (numerical or otherwise) have to be interpreted, and the mathematician gives him various mathematical models to that end. The scientist then selects the most suitable model and the values of parameters, predicting from the model what his experiment should show, and sees whether further experiments fulfil the prediction. If not, the model is modified or discarded. Nothing, or almost nothing, of this can be exact; exactness can normally only occur in pure mathematics; experimental error occurs in various forms. For example, in the amplification of varying electric currents a random noise is produced by the thermal motion of electrons in the amplifier, and this can often be avoided by using a maser operating at very low temperatures. Experimental error has as many forms as the number of kinds of experiments performed, and it increases as we pass from physics to chemistry and then biology. In subjects such as economics, anthropology, linguistics, one cannot experiment, one can only collect the data as they occur, and the error is usually greater than that in physical systems. The study of such data, and the errors of experiments, has produced two more branches of applied mathematics, those of statistics and statistical physics. In other branches it is usually vital to smooth or rationalize, to obtain a suitable pure mathematical model. If it gives correct predictions we replace the experimental results by the model. This occurs in statistical physics, the basic model being the theory of probability. But in statistics, though the model is roughly the same, we cannot discard the raw material and use the model as this is the source of many fallacies. We have to keep separate the raw material and the pure mathematical model, being repaid by a resulting greater precision of possible statements in the subject.

The first operation in statistics is that of *classification*. If the data consist of single real numbers x we can divide up their range into a finite number of disjoint sets for example, the sets given by

$$x < -n,\ -n \leq x < -n+1,\ -n+1 \leq x < -n+2, \ldots,\ 0 \leq x < 1, \ldots, n-1 \leq x < n,\ x \geq n,$$

to take a very simple case. If the data are a fixed number p of real numbers we can represent each value as a point of R^p and we can again divide up the space into a finite number of disjoint sets. In practice we cannot observe a complete infinite sequence of real numbers, but we may have data in which p can be arbitrarily large, for example in sequential sampling, in which case we use an infinite dimensional Cartesian space which can be divided up in a suitable way into a finite number of disjoint sets, by dividing the space only with respect to a finite number of co-ordinates and leaving the rest undivided. We can observe continuous functions, an early example being the traces produced by a barograph, and a more modern example being heart muscle action recorded on a cathode ray tube, and we need a space of continuous functions that can be divided up by, for example, taking a finite number of values of the independent variable or variables and classifying the values of the functions at those points. Some observations cannot be measured, or can only be measured at great inconvenience, for example, the grading by eye of colours or shades; here, classification becomes difficult.

After classification we are given a finite number, say, $S_1, S_2, \ldots, S_m$ of disjoint sets with union the whole set of possible data values, and the given finite number of data values from the experiment or other research. We count the number N_j of times that the data values fall into the set S_j, for $j = 1,2,\ldots,m$, the total number of values being N, the sum of the N_j. The relative frequency of values in the set S_j is defined to be $f_j^N = N_j/N$, a ratio lying between 0 and 1, the sum of the relative frequencies for $j = 1,\ldots,m$, being 1. If $j \neq k$, the relative frequency of values in $S_j \cup S_k$ is $f_j^N + f_k^N$, so that the relative frequency is finitely additive.

If we have made a proper classification and if we increase the amount of data, it usually happens that the f_j^N tend to fixed values. If they do not, it may be an accidental oscillation or it may be that an important source of variation has been neglected. For example, we might think that a process is independent of time, but the collected data might show a rise in time. In the pure mathematical model of the relative frequency we attach a number p_j, called the *probability of* S_j, to the set S_j ($j = 1, \ldots, m$), that lies between 0 and 1, the sum for $j = 1, \ldots, m$ being 1, such that the probability of $S_j \cup S_k$ is $p_j + p_k$ ($j \neq k$) so that the probability is finitely additive. The problem is to link p_j with f_j^N, an obvious way being to write $p_j = \lim_{N\to\infty} f_j^N$, but this cannot be given a rigid interpretation. First, in practice we cannot observe all terms of an infinite sequence without taking an infinite time over it. Secondly, even if we could, and if the limit exists for one set of data, nothing ensures that the limit will exist for another set of the same kind of data, or that the two limits are the same, except trivially when $f_j^N = 0$ for all N, or $f_j^N = 1$ for all N, and similar cases. von Mises defined a *Kollektiv* as an infinite sequence (x_j) of 0's and 1's with the properties that the relative frequency of 0's in the first N members of the sequence tends to a limit as N tends to infinity, and that the same is true, with the same limit, for an arbitrary subsequence of the (x_j). By taking the subsequence to consist of 1's alone, or of 0's alone, it is clear that the only Kollektivs are those sequences for which either $x_j = 0$ for all j greater than a fixed J, or $x_j = 1$ for all j greater than a fixed J. It was not the intention of von Mises to restrict his Kollektivs so drastically, so that some restriction has to be applied to the choice of subsequence. von Mises made the choice without knowing the values x_j, but the accurate definition of such ideas requires probability, so that the argument is circular.

Let us dig more deeply, writing each (x_j) as a decimal in the scale of 2 and between 0 and 1, $x = 0, x_1x_2\ldots x_j\ldots$. Each number in [0,1] gives a sequence (x_j), and different sequences (x_j), (y_j) can only give the same x when there is an integer J such that $x_j = y_j$ when $j < J$, while one sequence subsequently takes the values 0, and 1 repeated, the other taking the values 1, and 0 repeated. These x are multiples of powers of ½, and their set is countable and can be neglected. Now let constants m_1, m_2 have sum 1 and lie in (0,1). There is a continuous non-negative finitely additive interval function p satisfying

$$p(0,1) = 1, p(2m.2^{-j}, (2m+1)2^{-j}) = m_1 p(m.2^{-j+1}, (m+1)2^{-j+1}),$$

$$p((2m+1)2^{-j}, (2m+2)2^{-j}) = m_2 p(m.2^{-j+1}, (m+1)2^{-j+1}).$$

The set X_{MN} of all x with $x_1+x_2+\ldots+x_N = M$, is a finite union of intervals and

$$(20.1) \quad V(p;[0,1);X_{MN}) = {}^N C_M m_1^{N-M} m_2^M = w_M,$$

say. Given $\varepsilon > 0$, let k be the integer next lower than $Nm_2 + N\varepsilon$. Then for $j > k$

$$\frac{w_{j+1}}{w_j} = \frac{(N-j)m_2}{(j+1)m_1} < \frac{(N-Nm_2-N\varepsilon)m_2}{(Nm_2+N\varepsilon)m_1} = \frac{1-\varepsilon/m_1}{1+\varepsilon/m_2} = c, \; 0 < c < 1.$$

Thus if Y_N is the set where $x_1 + \ldots + x_N > m_2N + 2\varepsilon N$, we use (20.1) to obtain

$$V(p:[0,1);Y_N) = \sum_{j>m_2N+2\varepsilon N} w_j < w_{k+1}c^{\varepsilon N-1}/(1-c) < c^{\varepsilon N-1}/(1-c),$$

$$V(p;[0,1); \bigcup_{N\geq M} Y_N) < c^{\varepsilon M-1}/(1-c)^2.$$

Hence the set Z_ε where $\limsup_{N\to\infty} (x_1+x_2 + \ldots + x_N)/N > m_2 + 2\varepsilon$ satisfies $V(p;[0,1);Z_\varepsilon) = 0$. Similarly for $\liminf_{N\to\infty} (x_1 + x_2 + \ldots + x_N)/N < m_2-2\varepsilon$. Taking $\varepsilon = 1/m$, $m = 1,2,\ldots$, we see that, p-almost everywhere,

$$(20.2) \quad \lim_{N\to\infty} (x_1 + x_2 + \ldots + x_N)/N = m_2.$$

Changing m_2 in (0,1), we do not change the limit of a particular Kollektiv, but we change the sets of variation zero. There is no criterion depending on (x_j) alone, to tell us whether the corresponding x lies in the set of variation zero, or whether x lies in the complement of that set, the criterion must depend on the particular value of m_2 chosen, m_2 must be chosen first, and its value cannot be deduced from the behaviour of (x_j). This is a deeper reason why we cannot define probability as the limit of the relative frequency, unless we choose a special m_2 a priori. G.A. Barnard has suggested that for p we use the Haar measure, which means that here we take $m_2 = \frac{1}{2}$.

Another definition is that probability is the degree of rational belief, and by using various axioms a theory can be built up. But the resulting probability is rather subjective and there seems to be no method of evaluating probabilities in statistics.

Probability can also be defined as a pure mathematical measure with various properties, but divorced from numerical data.

Each of the above definitions seems to hold part but not all of the ethos of probability in statistics, so we use a blend of all three, assuming that the applied statistician's numerical work is correctly carried out, but perhaps wrongly interpreted. Here we concentrate on tests of significance. When considering a possible statistical experiment we usually assume that a certain effect will not occur, known as a null hypothesis. Using our rational belief or otherwise in the experiment, we assign a number δ in $[0,\frac{1}{2}]$, called the *level of significance*. If the null hypothesis is a basic belief we put $\delta = 0$ and have no need to do the experiment. If, however, the null hypothesis could be false, we take δ positive, with value such that we accept a situation in which we reject the hypothesis, when true, in a proportion δ of cases. The more costly a rejection is, the nearer we take δ to 0. Experiments in parapsychology have the smallest δ, we test whether we can obtain information of the natural world or the thoughts of an agent, or can move bodies by ways that do not use the known senses or sciences. Many would say that such is impossible, putting $\delta = 0$. Others, conceding that the effect might happen, would have to be thoroughly convinced first, and these put $\delta = 10^{-10}$ or some still smaller positive number.

Having chosen δ, we arrange the test procedure so that when the hypothesis is true we reject it in a proportion δ of cases, and accept it in a proportion $1-\delta$ of cases, both proportions being true in the long run. The meaning of "in the long run" is defined mathematically, using probability in the pure mathematical sense. The choices of a suitable probability measure and of the significance level, are known as the *problem of specification*. This is often governed by the knowledge of previous experiments, by some theoretical argument, or by some hypothesis to be tested.

We arrange to obtain a sample of m independent values $y_1,\dots,y_m$ of a variable from the experiment that has not yet begun. The integer m may be fixed or may itself be a statistical variable. Pure mathematics can partially but never wholly solve the problem of ensuring that the values are independent, there is still much in the design of the experiment that has to be left to the individual assessment of the applied statistician. The probability measure of the variable gives that of the sample, and we choose a set $C(\delta)$ of samples that has $1-\delta$ for probability measure, where δ is the significance level. If the variable is discrete this is not always possible, and then we sometimes use sets $C(\delta)$ with probability measure greater than $1-\delta$. For variable m, the $C(\delta)$ depends on m, and then we also use $C(\eta)$ for η in $(\delta,1)$ and near to 1, with $C(\delta) \supset C(\eta)$.

For fixed m we carry out the experiment and obtain the sample. If in $C(\delta)$, the sample is *compatible with the null hypothesis* and is *not significant*, relative to δ and $C(\delta)$. Otherwise the sample is *incompatible with the hypothesis*, and is *significant*, and the null hypothesis is rejected. For variable m, if the sample is in $C(\eta)$, the sample is *compatible with the hypothesis* and is *not significant*, relative to η and $C(\eta)$, and the null hypothesis is accepted. If the sample is not in $C(\delta)$ it is *incompatible with the hypothesis* and is *significant* relative to δ and $C(\delta)$, and the null hypothesis is rejected. If the sample is in $C(\delta)$ but not in $C(\eta)$ we take a further value y_{m+1} and retest. Sometimes there is a ceiling M beyond which m is not allowed to go.

The next idea is that of *independence*. If a variable x ranges over a set X and a variable y over Y, then (x,y) ranges over $X \times Y$. Supposing that there is a probability measure p_x over sets in X, a p_y over sets of Y, and p over sets of $X \times Y$, if

$$p_x(I)p_y(J) = p(I \times J)$$

for all bricks $I \subseteq X$, $J \subseteq Y$, we say that x and y are *independent*. Strictly, this is not a property of the variables x,y, but of the models p_x, p_y, p that we are using. This property is often tested by using contingency tables, and is assumed when we take a sample of m independent values. If for N independent samples the sample is in $C(\delta)$ exactly M times, we use independence and the property of combinations as in the binomial theorem, to show that the probability measure of this is ${}^{N}C_{M}(1-\delta)^{M}\delta^{N-M}$. Thus as in the proof of (20.2) we see that $M/N \to 1-\delta$ as $N \to \infty$, except for a set of samples of probability measure 0. This gives a pure mathematical definition of the statement that M/N tends to the right value in the long run, and it is the simplest of such convergence results, deeper results also being true.

There are many regions $C(\delta)$, so the question arises, which is the best? Our construction can be described by saying that we commit an *error of the first kind* when we reject the null hypothesis when it is true, and we fix the probability measure of the error at δ. We also commit an *error of the second kind* if we accept the null hypothesis when it is false, and we wish to minimize the probability measure of errors of the second kind. These two probability measures are not the same, the former being p, say, and the latter being an alternative probability measure, say q. Note that we can only test p against a single function q using these ideas, and so only one alternative hypothesis, unless several hypotheses give rise to the same q.

When p,q are continuous in one dimension, they are non-negative and clearly AC* with respect to p+q, and then by Theorem 16.5 (16.11), (16.12) there exist $f,g \geq 0$ such that

$$(20.3) \qquad p = \int fd(p+q), \quad q = \int gd(p+q).$$

In dimension $n > 1$ we have not proved a similar result, because of the rigid geometrical proof of Vitali's theorem, but we can assume the existence of f,g, as (20.3) can be proved in another way.

Theorem 20.1: *Let* S *be the set of all samples, supposed a finite or infinite brick, and* U(r) *the set of samples s satisfying*

$$(20.4) \qquad g(s) \leq rf(s),$$

for r *a real constant. If* U(r) *is measurable and if a measurable set* U *of* samples satisfies

$$V(p;S;U) = V(p;S;U(r)),\ V(p;S;U\setminus U(r)) > 0,$$

then

$$V(q;S;U) > V(q;S;U(r)).$$

Proof: Using (20.3), and the variations as integrals, if (20.4) holds for all $s \in U$, a measurable set, then

$$V(q;S;U) \leq rV(p;S;U).$$

If instead, for all $s \in U$, a measurable set,

$$(20.5) \quad g(s) > rf(s)$$

then either $V(q;S;U) > rV(p;S;U)$, or there is equality, in which case U has p-variation zero. For if X is the indicator of U,

$$\int_S X\cdot(g-rf)d(p+q) = 0,\ X\cdot(g-rf) = 0$$

(p+q)-almost everywhere and

$$0 \leq V(p;S;U) \leq V(p+q;S;U) = 0.$$

Now sets $U(r)\setminus U$ and $U\setminus U(r)$ are measurable and are in regions where s satisfies (20.4) and (20.5), respectively. Hence

$$V(q;S;U\setminus U(r)) > rV(p;S;U\setminus U(r)) = rV(p;S;U) - rV(p;S;U \cap U(r))$$

$$= rV(p;S;U(r)) - rV(p;S;U \cap U(r)) = rV(p;S;U(r)\setminus U) \geq V(q;S;U(r)\setminus U),$$

$$V(q;S;U) = V(q;S;U \cap U(r)) + V(q;S;U\setminus U(r)) > V(q;S;U \cap U(r)) + V(q;S;U(r)\setminus U)$$

$$= V(q;S;U(r)).$$

Thus if 1-V(p;S;U(r)) is used as a significance level, U(r) is the best possible set to distinguish between p and q. If g = rf for some constant r and in some measurable set U of positive p-variation, then V(p;S;U(r)) is discontinuous at that r. Otherwise we can find an r to give the required significance level.

If there is more than one alternative probability measure q, the best possible region U(r) for one q might not be the best for another q, so that we have to make a compromise. But here we do not go further into such details.

Various integrals are associated with the probability measure p of a variable x, of the form $\int_T f(x)dp$, where T is the finite or infinite brick over which x ranges. f(x) = x gives the *mean* μ, and for $f(x) = (x-\mu)^2$ we have the *variance*, the square root being the *standard deviation*. For a constant t, $f(x) = x^t$ gives the t*th. moment*, and for various values of t, the *generating function of* p. $f(x) = e^{xt}$ gives the *moment generating function*, and $f(x) = e^{ixt}$ the *characteristic function*, which is nothing to do with the characteristic function of a set.

Let x,y be independent variables on the line with continuous probability measures p,q respectively, and let X(t) be the measurable set of (x,y) with x+y ≤ t, so x ≤ t-y. If P is the whole plane the probability measure of X(t), and so of x+y as t varies, is

$$\int_P \chi(X(t);(x,y))d(pq)$$

where $\chi(X(t);(x,y))$ is the indicator of X(t). By Fubini's theorem, Theorem 17.2, and the continuity of p, this integral is equal to

$$\int_{(-\infty,\infty)}\left(\int_{(-\infty,\infty)} \chi(X(t);(x,y))dp\right)dq = \int_{(-\infty,\infty)} p(t-y)dq,$$

the *convolution of* p *and* q. The way is now open to the Central Limit and allied theorems. Next we look at Karl Pearson's *correlation ratio* η_{yx} connected with the dependence of two variables x,y. The probability measure form is as follows.

Let the probability measure associated with points z = (x,y) of the plane P be obtained from $p = \int f(z)dA$, A being the area function, to avoid confusion with the mean. We write

$$g(x) \equiv \int_{(-\infty,\infty)} f(x,y)dy > 0, \ \bar{y}(x) = \int_{(-\infty,\infty)} yf(x,y)dy/g(x),$$

$$\bar{y} = \int_P yf(z)dA,$$

$$(20.6) \quad 1-\eta_{yx}^2 \equiv \int_P (y-\bar{y}(x))^2 f(z)dA / \int_P (y-\bar{y})^2 f(z)dA.$$

$$\int_P (y-\bar{y}(x))^2 f(z)dA = \int_P (y-\bar{y})^2 fdA + 2\int_P (y-\bar{y})(\bar{y}-\bar{y}(x))fdA + \int_P (\bar{y}-\bar{y}(x))^2 fdA$$

$$= \int_P (y-\bar{y})^2 fdA + 2\int_{(-\infty,\infty)} (\bar{y}-\bar{y}(x))(g(x)\bar{y}(x)-g(x)\bar{y})dx + \int_{(-\infty,\infty)} (\bar{y}-\bar{y}(x))^2 g(x)dx$$

$$= \int_P (y-\bar{y})^2 fdA - \int_{(-\infty,\infty)} (\bar{y}-\bar{y}(x))^2 g(x)dx,$$

$$(20.7) \quad \eta_{yx}^2 = \int_{(-\infty,\infty)} (\bar{y}-\bar{y}(x))^2 g(x)dx / \int_P (y-\bar{y})^2 f(x,y)dA.$$

Thus from (20.6), (20.7) we can prove the following respective results.

<u>Theorem 20.2</u>: *If* $\eta_{yx} = \pm 1$ *then* $y = \bar{y}(x)$ *except in a set* Z *of* (x,y) *with*

$$V(p;P;Z) = V(fA;P;Z) = V(A;P;Z \cap Z_0) = 0.$$

where Z_0 *is the set where* $f \neq 0$. *If* $\eta_{yx} = 0$ *then* $\bar{y}(x) = \bar{y}$ *except in a set* X *of* x *with*

$$V(\int gdx;(-\infty,\infty);X) = V(g\Delta x;(-\infty,\infty);X) = V(\int_{(-\infty,\infty)} f(x,y)dy\Delta x;$$
$$(-\infty,\infty);X) = 0.$$

Thus, apart from sets of probability measure 0, there is in the first case a complete functional dependence of y on x, while in the second case the mean of y for each x does not vary with x. In the very special case when both are true, so that y is almost everywhere equal to a constant b, then $\bar{y} = b$ also and

$$\int_P (y-\bar{y})^2 f(x,y)dA = 0.$$

Finally, in $T = R^n$ let $x = (x_1,\ldots,x_n)$, let $T(x)$ denote the infinite brick consisting of all $w = (w_1,\ldots,w_n)$ satisfying $w_j < x_j$ ($j = 1,\ldots,n$) and let p be a probability measure on R^n. Then $F(x) \equiv p(T(x))$ is called the *distribution function of* p. Let $G(t,x)$ be a distribution function in $t = (t_1,\ldots,t_m) \in R^m$ with probability measure $g(x,I)$ ($I \subseteq R^m$), for p-almost all x, and for every t let $G(t,x)$ be a p-measurable function of x. Then the function

$$H(t) = \int_T G(t,x)dF(x)$$

is a distribution function in R^m, with probability measure q, say, and

$$q(I) = \int_T g(x;I)dF(x).$$

These follow from Fubini's theorem (Theorem 17.2).

CHAPTER 8
HISTORY AND FURTHER DISCUSSION

21 Other Integrals

At the beginning of the book, Section 1 gives a brief introduction to the origins of integration and the integral of Cauchy, Riemann, and Darboux, followed in Section 2 by those of Stieltjes, J.C. Burkill, and Moore and Smith (refinement integral). Then Section 3 goes back to the calculus (Newton) indefinite integral, and forward to the modern generalized Riemann integral. This last includes the previously mentioned integrals. Those in Section 1, 2 that are not refinement integrals are defined by δ-fine divisions with δ a positive constant. For refinement integrals see Theorem 4.6 Corollary. By (3.16) the calculus integral is included. In time, Riemann's integral was followed by Lebesgue's, and Theorem 3.4 shows that it is also included. The inclusion of the calculus and Lebesgue integrals make it not surprising that the first integral to include these, the special Denjoy integral, is itself included in the generalized Riemann integral. This follows since the Cauchy extension of integrals included in generalized Riemann integrals and over expanding intervals or bricks, is also included, Theorem 5.14, while the Harnack extension of generalized Riemann integrals is a generalized Riemann integral (Theorem 5.15). These two extensions were applied by Denjoy first to Lebesgue integrals and then to the integrals so found, in a transfinite inductive process. Thus at every step the integral is a generalized Riemann integral. Denjoy proved that in integrating derivatives the process terminates after an at most countable number of steps.

O. Perron (1880-1975) defined an integral in 1914 using derivates in one dimension that is easily shown to be equivalent to the variational integral of the interval-point function of the type $f(x)\,\Delta x$, which in turn is equivalent to the corresponding generalized Riemann integral (Theorem 5.3 and following definitions). As H. Hake proved in 1921 that the special Denjoy integral is included in the Perron integral and as P. Alexandroff in 1924 and H. Looman independently in 1925 proved the reverse inclusion, we have a second proof of the equivalence of the Denjoy special integral of f with the generalized Riemann integral of $f\Delta x$. See Theorems 21.1, 21.2.

In 1912, N. Lusin (41) gave a descriptive definition of the Denjoy integral of f over [a,b]. It is ΔF where F is ACG* and has f as derivative almost everywhere, and in recent years several have proved that Lusin's integral is equivalent to the generalized Riemann integral of $f(x)\Delta x$ in one dimension, which give other proofs of equivalence. See Yôto Kubota (31), and Lee Peng-Yee and Wittaya Naak-in (36), for example. Here we define the Perron integral in the following way. For $[a,b] \subseteq R^1$, b-a finite, let f,M be functions from [a,b] to R^1. Then the lower derivate $\underline{M}'$ of M is the lower limit of $\Delta M/\Delta x$ as the interval shrinks round a fixed point, with appropriate one-sidedness at a and b. M is defined a Perron major function of f in [a,b] if M is continuous on [a,b] with $M(a) = 0$, $\underline{M}' \geq f$ almost everywhere on [a,b], and $\underline{M}' > -\infty$ except for a countable set X on [a,b].

This definition is in a sense an amalgam of the worst features of the definitions of O. Perron (51), H. Bauer (2), H. Hake (17), H. Looman (40), S. Saks (54), pp. 186-203, and E.J. McShane (43), p. 313, except that McShane used two functions like $\underline{M}'$, and that our f is always finite-valued. The last causes no trouble since if a function is Perron integrable it is finite almost everywhere, we can replace f by 0 in the exceptional set, and we can include it with the previous set of variation zero.

The infimum of M(b) for all Perron major functions M of f on [a,b], is called the upper Perron integral of f on [a,b]. An $m:[a,b] \to R^1$ is a Perron minor function of f on [a,b], if and only if -m is a Perron major function of -f on [a,b]. The lower Perron integral of f on [a,b] is the supremum of m(b) for all such m, and if the upper and lower Perron integrals are equal, their value is the Perron integral of f on [a,b]. Further, we say that M is a strong Perron major function of f on [a,b], if and only if M is continuous on [a,b] with $M(a) = 0$ and $\underline{M}' > f$ everywhere on [a,b], while m is a strong Perron minor function on f on [a,b] if and only if -m is a strong Perron major function of -f on [a,b].

<u>Theorem 21.1</u>: The upper and lower Perron integrals of f on [a,b] are unaltered if the Perron major and minor functions of f are restricted to

strong Perron major and strong Perron minor functions of f, respectively, and in consequence, if f is replaced by an $f^* : [a,b] \to R^1$ with $f^* = f$ almost everywhere.

Proof: Using the continuity of a Perron major function M of f we remove the sequence (x_j') of points where $\underline{M}' = -\infty$. Let

$$L(x,H) = \sup_{0<|h|\leq H} |M(x+h)-M(x)|, \; L(x,0) = 0, \; L(x,-H) = -L(x,H).$$

Then for each fixed x, L(x,H) is bounded and continuous in H and monotone decreasing to 0 as $H \to 0+$. Thus, given $\varepsilon > 0$, there are an $H_j > 0$ and $L_j(x)$ such that

$$(21.1) \quad L(x_j,H_j) < \varepsilon.2^{-j-1} \quad (j = 1,2,\dots),$$

$$(21.2) \quad L_j(x_j+x) = \begin{cases} L(x_j,x) - L(x_j,-H_j) & (|x| \leq H_j), \\ L(x_j,H_j) - L(x_j,-H_j) & (x \geq H_j), \\ 0 & (x \leq -H_j). \end{cases}$$

Then for $0 < h \leq H_j$, using (21.1), (21.2),

$$L_j(x_j+h)-L_j(x_j) \geq |M(x_j+h) - M(x_j)|, \; L_j(x_j)-L_j(x_j-h) \geq |M(x_j)-M(x_j-h|),$$

$$0 \leq \sum_{j=1}^{\infty} L_j \leq \varepsilon, \text{ and } M_1 \equiv M + \sum_{j=1}^{\infty} L_j$$

is continuous in [a,b] with $\underline{M}_1' \geq 0$ at each x_j. As a countable set is of measure or variation zero, M_1 is a Perron major function of f with X empty.

Next we remove the exceptional set Y of measure or variation zero where $\underline{M}' < f$. For each j there is an open set $G_j \supseteq Y$ of measure or variation less than $\varepsilon.4^{-j}$. We define

$$y_j(x) = \begin{cases} 2^j & (x \in G_j), \\ 0 & (x \in \sim G_j), \end{cases} \qquad z_j(x) = \int_{[a,x)} y_j(x)dx, z(x) \equiv \sum_{j=1}^{\infty} z_j(x), z_j(b) < \varepsilon 2^{-j},$$

$z(b) < \varepsilon$.

Then z_j and z are continuous and monotone increasing in $[a,b]$. Putting

$$M_2(x) \equiv M_1(x) + z(x) + \varepsilon(x-a), \underline{M}_1' > -\infty$$

gives

$$\underline{M}_2' = -\infty \ (x \in Y), \ \underline{M}_2' \geq f + \varepsilon > f \ (x \in [a,b]\setminus Y).$$

Thus the continuous M_2 is a strong Perron major function of f on $[a,b]$ and Theorem 21.1 follows from $M(b) < M_2(b) < M(b) + \varepsilon(2+b-a)$.

Theorem 21.2: *If* M,m *are strong Perron major and strong minor functions, respectively, of* f *on* $[a,b]$ *then* $M-m$ *is monotone increasing. Further, if the Perron integral exists over* $[a,b]$ *it exists, say as* $P(t)$*, over* $[a,t]$*, for* $a < t \leq b$*, and if* $P(a) = 0$ *then* P *is the variational integral of* f *over* $[a,b)$*. Conversely, if the variational integral exists, so does the Perron integral, and the two are equal.*

Proof: By Borel's covering theorem and using in $[u,v] \subseteq [a,b]$, and $\bar{m}' = -(-\underline{m})'$,

$$\bar{m}' < f < \underline{M}', \ \underline{(M-m)}' > 0, \ M(v)-m(v)-m(u)+m(u) > 0,$$

and $M-m$ is strictly increasing for each such m,M, and the upper integral over $[a,b]$ is not less than the lower integral. If the Perron integral exists over $[a,b]$ then $M(b)-m(b) < \varepsilon$, for suitable m,M, given $\varepsilon > 0$. As $M(a) = 0 = m(a)$ with $M-m$ strictly increasing, $M(t)-m(t) < \varepsilon \ (a < t \leq b)$, and as $\varepsilon > 0$ is arbitrary the Perron integral exists over $[a,t]$. Write it as $P(t)$ with $P(a) = 0$, and let m tend to P. Then M-P is monotone increasing and non-negative, and similarly so is P-m. Then from $\underline{M}' > f$ and all intervals in the neighbourhood of a fixed end-point, we have

$$\Delta M > f\Delta x,\ \Delta M-\Delta P \geq f\Delta x - \Delta P,\ \Delta m-\Delta P \leq f(x)\Delta x - \Delta P,$$

$$|f\Delta x - \Delta P| \leq (\Delta M-\Delta P) + (\Delta P-\Delta m) = \Delta(M-m),\ (\mathcal{D})\ \Sigma|f\Delta x-\Delta P| \leq (\mathcal{D})\ \Sigma\Delta(M-m)$$

$$= M(b) - m(b).$$

Given $\varepsilon > 0$, by choice of M,m, this difference is less than ε. Hence P is the variational integral of f over [a,b]. Conversely, if $P(t)$ is the variational integral of f over $[a,t)$ $(a < t \leq b)$, then using Theorem 5.3 (5.6) with $h = f\Delta x$, $H = \Delta P$, and ε for 8ε, and a suitable positive function δ,

$$f\Delta x \leq |f\Delta x-\Delta P| + \Delta P \leq V(f\Delta x-\Delta P;\Delta x) + \Delta P \equiv \Delta M,\ f\Delta x \geq \Delta P-V(f\Delta x-\Delta P;\Delta x) \equiv \Delta m,$$

$$\bar{m}' \leq f \leq \underline{M}',\ V(f\Delta x-\Delta P;\ [a,t))$$

being monotone increasing in t. Hence M and m are Perron major and minor functions, respectively, of f over [a,b], and as the variation is as small as we please, P is also the Perron integral of f.

This section shows that several integrals are included in the generalized Riemann integral in Euclidean space of a finite number of dimensions. However, the methods used have a far greater scope than is shown in this book. The original theory was based on a battery of axioms, but in preparation of the book, Henstock (27), J.J. McGrotty suggested the use of something to parallel the measure spaces of Lebesgue theory. So I introduced division spaces, see Chapter 10, and particularly the examples given on pp. 219 to 225. Since then, the number of examples of integrals included in the division space theory has grown considerably, covering practically all integrals that do not use convergence factors (smoothing devices) in their definition. Convergence-factor integrals are included in the system of N-variational integrals. For division space theory see, for example, R. Henstock (27), P. Muldowney (46), K.M. Ostaszewski (49), and references contained therein.

For numerical work on integration the simple and dominated integrals were introduced in Haber and Shisha (14), (15), and Osgood and Shisha (47), (48), respectively. Lewis and Shisha (39) show that these are included in the integral of this book.

22 Notes On The Previous Sections

In this final section our task is to show the genesis and development of the various proofs and other details in the previous sections. Section 1 and part of Section 2 contain in themselves the required history. However, Theorems 2.2, 2.4 seem new, while Theorem 2.5 comes from J.C. Burkill (6). Theorems 2.7, 2.8, 2.9 are developments of W.H. Young (58). Theorem 2.10 and corollary are a development of E. Helly (19). Theorems 2.12, 2.13 are obvious and Lemma 2.14 is well known. Theorems 2.15, 2.16 are due to G.H. Hardy (18), p. 187, and correspond to the Abel and Dirichlet tests of convergence of series, and similarly for Theorem 2.18. Theorem 2.19 is built on the well known integral test, while for more than Theorem 2.20 one may look at A.M. Ostrowski (50).

In Section 3 we begin with the ordinary calculus and progress to the generalized Riemann integral that uses (3.17). After many attempts to give a Riemann definition to Lebesgue integration, see H. Lebesgue (34), pp. 30-33, E. Borel (3), (4), H. Hahn (16), A. Denjoy (11), (12), B. Levi (38), the first systematic efforts seem to be in J. Kurzweil (32), and in R. Henstock (22), pp. 277-278, (23), and (24). On 3rd October, 1963, K. Kartâk informed me of J. Kurzweil's paper, so that until then, Kurzweil and myself were working independently. Theorem 3.1 was first given in P. Cousin (8), p. 22, for two dimensions. Lusin (42) used the idea for trigonometrical series, while W.H. and G.C. Young (59) state the result. The first incomplete proofs seem to be in J. Kurzweil (32), p. 423, Lemma 1.1.1, and R. Henstock (22). Theorem 3.4 is taken from R.O. Davies and Z. Schuss (10) while Theorem 3.5 is from R. Henstock (24) pp. 34-35, Theorem 22.2. Theorem 4.1 is the n-dimensional case of Theorem 3.1, and some further theorems are obvious. Note that complex-valued functions are integrated by the same definition as real-valued functions, and Theorem 4.5

provides a link. The proof of Theorem 4.6 (4.7) can be improved by proving (4.8) first, while for Theorem 4.7 see R. Henstock (24), p. 34, Theorem 22.1, for Exx. 4.5, 4.6, 4.8, 4.9, see the same book, p. 29, Exx. 20.1, 20.2, and for Theorem 5.1 see the book, p. 31, Theorem 21.1. Some call the inequality (5.5), Henstock's lemma. But I copied a proof of S. Saks (53) for Burkill integration, see R. Henstock (20), (21). My contribution was to give the proof in generalized Riemann integration and to use it in proofs of limit theorems. Thus a better name is the "Saks-Henstock lemma". The neat proof of Theorem 5.1 comes from R. Henstock (29), p. 401, Theorem 1. For Theorem 5.3 see the book, p. 40, Theorem 24.1. Theorem 5.4 coalesces several similar results, e.g. the book, p. 57 Theorem 30.1. For Theorem 5.6 see p. 49, Theorem 28.1. For Theorem 5.11 see the book, pp. 54-55, Theorem 28.6. For integration by parts, Theorem 5.12, see the book, p. 69, Theorem 33.1. The Cauchy extension, Theorem 5.14, is in the book, pp. 113-115, Section 46, and the extension to integration over infinite intervals is in the book, pp. 115-118, Section 47. The Harnack extension, Theorem, 5.15, is called the Denjoy extension in the book, pp. 118-120, Section 48. Section 5 is an important fundamental section, so that its origins need careful documentation.

Section 6 originated from the book, pp. 43-46, Section 25, improved by R. Henstock (29). Section 7 on the variation set originated from the book, p. 32-33, but has developed considerably since then.

Section 8, the first on limits under the integral sign, began in the book, pp. 82-85, Section 36, while Section 9 began with R. Henstock (28), p. 528, Theorem 18, blended with the book, pp. 85-87, Section 37. Section 10 on controlled convergence is taken from P.Y. Lee and T.S. Chew (35), Section 11 is completely new, and Section 12 has much in common with the book, pp. 127-140, Chapter 9, E. Hölder (30), H. Minkowski (44), and W.H. Young (58). Section 13 translates many results into results for more general convergence, while Section 14 looks at Fatou's lemma, following R. Henstock (29). Section 15 applies the previous Section 13 to differation of integrals, with W. Sierpinski's (55) and Vitali's (57) covering theorems, and Section 16 updates the book, pp. 92-98 Section 39. There have been many improvements in the proof of Fubini's Theorem 17.2

since the proof in R. Henstock (23). Theorem 17.3 is due to T.W. Lee (37). Sections 18, 19 are new, while Section 20 is a condensation of the book, pp. 148-161.

REFERENCES

(1) Alexandroff, P., 'Über die Äquivalenz des Perronschen und des Denjoyschen Integralbegriffes', Math. Zeitschrift 20, 213-222 (1924).

(2) Bauer, H., 'Der Perronsche Integralbegriff und seine Beziehung zum Lebegueschen,' Monatsh. Math. Phys. 26, 153-198 (1915).

(3) Borel, E., 'Sur la définition de l'intégrale définie', Comptes Rendus 150, 375-378 (1910).

(4) Borel, E., 'Sur une condition générale de l'intégrabilité', Comptes Rendus 150, 508-510, (1910).

(5) Burkill, J.C., 'Functions of intervals', Proceedings London Math. Soc. (2) 22, 275-310, (1924).

(6) Burkill, J.C., 'Differential properties of the Young-Stieltjes integrals', Journal London Math. Soc. 23, 22-28 (1948), Math. Reviews 10-185.

(7) Cauchy, A.L., Cours d'analyse de l'École Royale Polytechnique, 1re partie; analyse algebrique (1821, Paris), Works (2) 3.

(8) Cousin, P., 'Sur les fonctions de n variables complexes', Acta Math. 19, 1-.... (1894).

(9) Darboux, J.G. 'Memoire sur les fonctions discontinues', Ann. Sci. Ec. Norm. Sup. (2) 4.57-112, (1875).

(10) Davies, R.O., and Schuss, Z., 'A proof that Henstock's integral includes Lebesgue's', Journal London Math. Soc. (2) 2, 561-562 (1970), Math. Reviews 42 #435.

(11) Denjoy, A., 'Sur intégration riemanienne', Comptes Rendus Acad. Sci. Paris 169, 219-220 (1919).

(12) Denjoy, A., 'Sur la définition riemannienne de l'intégrale de Lebesgue', Comptes Rendus Acad. Sci. Paris 193, 695-698 (1931), Zent. für Math. 3.106.

(13) Dollard, J.D., and Friedman, C.N., Product Integration (Addison-Wesley, 1979) Math. Reviews 81e:34003.

(14) Haber, S., and Shisha, O., 'An integral related to numerical integration', Bulletin American Math. Soc. 79 930-932 (1973), Math. Reviews 47 #7288.

(15) Haber, S., and Shisha, O., 'Improper integrals, simple integrals, and numerical quadrature', Journal Approximation Theory 11, 1-15 (1974), Math. Reviews 50 #5309.

(16) Hahn, H., 'Über annaherung der Lebesgueschen integrale durch Riemannsche summen', Sitzber, Akad. Wiss Wien Abt. 123, 713-743 (1914).

(17) Hake, H., 'Ueber de la Vallee Poussins Ober- und Unterfunktionen einfacher Integrale und die Integraldefinition von Perron', Math. Annalen 83, 119-142 (1921)

(18) Hardy, G.H., 'Notes on some points in the integral calculus (I) On the formula for integration by parts', Messenger of Math. 30, 185-187 (1901).

(19) Helly, E., Über lineare Funktionaloperationen', Sitzungsberichte der Naturwissenschaftlichen Klasse der Kaiserlichen Akademie der Wissenschaften 121, 265-297 (1921).

(20) Henstock, R., 'On interval functions and their integrals', Journal London Math. Soc. 21, 204-209 (1946), Math. Reviews 8, 572.

(21) Henstock, R., 'On interval functions and their integrals (II)', Journal London Math. Soc. 23, 118-128 (1948), Math. Reviews 10, 239.

(22) Henstock, R., 'The efficiency of convergence factors for functions of a continuous real variable', Journal London Math. Soc. 30, 273-286 (1955), Math. Reviews 17, 359.

(23) Henstock, R., 'Definitions of Riemann type of the variational integrals', Proceedings London Math. Soc. (3) 11, 402-418 (1961), Math. Reviews 24 #A1994.

(24) Henstock, R., Theory of Integration (Butterworths, London, 1963), Math. Reviews 28 #1274.

(25) Henstock, R., 'The integrability of functions of interval functions', Journal London Math. Soc. 39, 589-597 (1964), Math. Reviews 29, #5975.

(26) Henstock, R., 'A Riemann-type integral of Lebesgue power', Canadian Journal of Math. 20, 79-87 (1968), Math. Reviews 36 #2754.

(27) Henstock, R., Linear Analysis (Butterworths, London, 1968), Math. Reviews 54 #7725.

(28) Henstock, R., 'Generalized integrals of vector-valued functions', Proceedings London Math. Soc. (3) 19, 509-536 (1969), Math. Reviews 40 #4420.

(29) Henstock, R., 'Generalized Riemann integration and an intrinsic topology', Canadian Journal of Maths. 32, 395-413 (1980), Math. Reviews 82b: 26010.

(30) Hölder, E., 'Über einen Mittelwertsatz', Nachr. Akad. Wiss. Göttingen (Math. Phys.) 38-47 (1889).

(31) Kubota, Y., 'A direct proof that the RC-integral is equivalent to the D*-integral', Proceedings American Math. Soc. 80, 293-296 (1980), Math. Reviews 81h:26006.

(32) Kurzweil, J., 'Generalized ordinary differential equations and continuous dependence on a parameter', Czechoslovak Math. Journal 7 (82) 418-446 (1957) (especially 422-428), Math. Reviews 22 #2735.

(33) Lebesgue, H., 'Intégrale, longueur, aire,' Annali Mat. Pura Appl. (3) 7.231-359 (1902), Jbuch 33, 307.

(34) Lebesgue, H., 'Sur les intégrales singulières,' Annls. Fac. Sci. Univ. Toulouse (3) 1, 25-117 (1909).

(35) Lee, P.Y. and Chew, T.S., 'A better convergence theorem for Henstock integrals', Bulletin London Math. Soc. 17 557-564 (1985), Math. Reviews 87b:26010.

(36) Lee, P.Y., and Wittaya, N., 'A direct proof that Henstock and Denjoy integrals are equivalent', Bulletin Malaysian Math. Soc. (2) 5, 43-47 (1982), Math. Reviews 84f:26011.

(37) Lee, T.W., 'On an extension of Fubini's theorem', Journal London Math. Soc. (2) 4, 519-522 (1972), Math. Reviews 46 #315.

(38) Levi, B., 'Teoria de la Integral de Lebesgue independiente de la nocion de Medida', Publ. Inst. Mat. Univ. Nac. Litoral 3, 65-116 (1941), Math. Reviews 3, 227.

(39) Lewis, J.T., and Shisha, O., 'The generalized Riemann, simple, dominated and improper integrals', Journal of Approximation Theory 38, 192-199 (1983), Math. Reviews 84h:26014.

(40) Looman, H., 'Ueber die Perronsche Integraldefinition', Math. Annalen 93, 153-156 (1925).

(41) Lusin, N., 'Sur les propriétés de l'intégrale de M. Denjoy', Comptes Rendus Acad. Sci. Paris 155, 1475-1478 (1912).

(42) Lusin, N., Integrals and trigonometric series (in Russian), Moscow, 1915.

(43) McShane, E.J., Integration (Princeton University Press, 1944), Math. Reviews 6, 43.

(44) Minkowski, H., Geometrie der Zahlen, I, 115-117 (1896).

(45) Moore, E.H., 'Definition of limit in general integral analysis', Proc. Natn. Acad. Sci. U.S.A., 1, 628-632 (1915).

(46) Muldowney, P., A general theory of integration in function spaces, Pitman Research Notes in Maths. 153 (Longmans, 1987).

(47) Osgood, C.F., and Shisha, O., 'The dominated integral', Journal of Approximation Theory 17, 150-165 (1976), Math. Reviews 54 #6467.

(48) Osgood, C.F., and Shisha, O., 'Numerical quadrature of improper integrals and the dominated integral', Journal of Approximation Theory 20, 139-152 (1977), Math. Reviews 56, #7128.

(49) Ostaszewski, K.M., 'Henstock integration in the plane', Memoirs American Math. Soc. 63 no. 353 (1986), Math. Reviews 87j:26016.

(50) Ostrowski, A.M., 'On Cauchy-Frullani integrals', Comment. Math. Helvetici 51, 57-91 (1976), Math. Reviews 53 #8347.

(51) Perron, O., 'Ueber den Integralbegriff', Sitzber. -B.Heidelberg Akad. Wiss., Abt. A 16, 1-16 (1914).

(52) Riemann, G.F.B., 'Über die Darstellbarkeit einer Function durch eine trigonometrische Reihe', Abh. Gesell. Wiss. Göttingen 13, math. kl. 87-132 (1868), Oeuvres mathematiques de Riemann (Ed. L. Laugel) (1898, Paris: reprinted 1968, Paris and Cleveland). Math. Reviews 36 #4952.

(53) Saks, S., 'Sur les fonctions d'intervalle', Fundamenta Math. 10 211-224 (p. 214) (1927), Jbuch 53, 233.

(54) Saks, S., Theory of the Integral, 2nd. English edition, Warsaw, 1937, Zent. fur Math. 17, 300.

(55) Sierpinski, W., 'Un lemme métrique', Fundamenta Math. 4, 201-203 (1923).

(56) Stieltjes, T.J., 'Recherches sur les fractions continues', Annales de la Faculte des Sciences de Toulouse 8, 1-122 (1894).

(57) Vitali, G., 'Sui gruppi di punti e sulle funzioni di variabili reali', Atti Accad. Sci. Torino 43, 75-92 (1908).

(58) Young, W.H., 'On classes of summable functions and their Fourier series', Proc. Roy. Soc. series A.87, 225-229 (1912).

(59) Young, W.H. and Young, G.C., 'On the reduction of sets of intervals', Proceedings London Math. Soc. (2) 14, 111-130 (1915).

Cameron, R.H. and Martin, W.T., 'An unsymmetrical Fubini theorem', Bulletin American Math. Soc. 47, 121-125 (1941), Math. Reviews 2; 257.

Denjoy, A., 'Une extension de l'intégrale de M. Lebesgue', C.R. Acad. Sci. Paris 154, 854-862 (1912).

Denjoy, A., 'Calcul de la primitive de la fonction dérivée la plus générale', C.R. Acad. Sci. Paris, Ser. A.B, 154, 1075-1078 (1912).

Harnack, A., 'Die allgemeinen Sätze über den zusammenhang der Funktionen einer reelen Variabeln mit ihren Ableitungen, II, Math. Annalen 24, 217-252 (1884).

McShane, E.J., Unified Integration (Academic Press, 1983), Math. Reviews 86c:28002

Neyman, J., and Pearson, E.S., 'On the problem of the most efficient tests of statistical hypotheses', Phil. Trans. A231, 289-337 (1933), Zent. für math. 6, 268.

Robbins, H., 'Mixture of two distributions', Annals of Math. Statistics 19, 360-369 (1948), Math. Reviews 10, 103.

de la Vallée Poussin, C., 'Étude des intégrales a limites infinies pour lesquelles la fonction sous le signe est continue', Annales de la Société Scientifique de Bruxelles, 16, 150-180 (1892).

INDEX

LECTURES ON THE THEORY OF INTEGRATION